Environments in a changing world

Environments in a changing world

John Huckle and Adrian Martin

An imprint of **Pearson Education**

Harlow, England · London · New York · Reading, Massachusetts · San Francisco · Toronto · Don Mills, Ontario · Sydney
Tokyo · Singapore · Hong Kong · Seoul · Taipei · Cape Town · Madrid · Mexico City · Amsterdam · Munich · Paris · Milan

Pearson Education Ltd
Edinburgh Gate
Harlow
Essex CM20 2JE
England

and Associated Companies around the World.

Visit us on the World Wide Web at:
www.pearsoneduc.com

First edition 2001

© Pearson Education Limited 2001

The rights of John Huckle and Adrian Martin to be identified as the authors of this
Work have been asserted by them in accordance with the Copyright, Designs and
Patents Act 1988.

ISBN 0582-32772-5

British Library Cataloguing-in-Publication Data
A catalogue record for this book can be obtained from the British Library

Library of Congress Cataloging-in-Publication Data

Huckle, John.
   Environments in a changing world / John Huckle and Adrian Martin. -- 1st ed.
     p. cm.
   Includes bibliographical references and index.
   ISBN 0-582-32772-5
     1. Human ecology.   I. Martin, Adrian.   II. Title.
   GF41 .H86 2001
   304.2--dc21

                                                              00-066894

10 9 8 7 6 5 4 3 2 1
05 04 03 02 01

Typeset in 11/12pt Adobe Garamond by 35
Produced by Pearson Education Malaysia Sdn. Bhd.,
Printed in Malaysia, VVP

Contents

Contents ix

Abbreviations

AOSIS	Association of Small Island States
BNFL	British Nuclear Fuels Limited
CIA	Central Intelligence Agency (United States)
CSE	Centre for Science and Environment (New Delhi)
DTER	Department of Transport, Environment and the Regions
EU	European Union
GCC	Global Climate Coalition
GM	genetically modified [crop]
GMO	genetically modified organism
GNP	gross national product
ICLEI	International Council for Local Environmental Initiatives
IMF	International Monetary Fund
IPCC	Intergovernmental Panel on Climate Change
IUCN	International Union for the Conservation of Nature
JFM	Joint Forest Management
LA21	Local Agenda 21
LETS	Local Economic Trading Scheme
LGMB	Local Government Management Board
MAI	Multilateral Agreement on Investment
NGO	non-governmental organization
NSM	new social movement
OECD	Organisation for Economic Cooperation and Development
UN	United Nations
UNEP	United Nations Environment Programme
UNESCO	United Nations Educational, Scientific and Cultural Organization
UNFCCC	United Nations Framework Convention on Climate Change
TCPA	Town and Country Planning Association
VFC	village forest committee
WCED	World Commission on Environment and Development
WFP	Western Forest Products
WRI	World Resources Institute
WTO	World Trade Organization
WWF	World Wide Fund for Nature

Acknowledgements

We are grateful to the following for permission to reproduce copyright material:

Figure 1.1 from The City As a Hybrid: On Nature, Society and Cyborg Urbanization in *Capitalism Nature So*, Vol. 7 No. 2 (Swynegedouw, E., June 1996); Figure 1.2 from *Political Ecology, Global and Local*, Roger Keil *et al.* (eds) published by Routledge, London (Altvater, E., 1998); Figure 1.5 from *Wealth Beyond Measure, an atlas of the new economics*, by Paul Ekins, Mayer Hillman and Robert Hutchinson, published by Gaia Books Ltd 1992, ISBN 1-85675-050-7; Figure 1.6, Figure 6.2, Figure 12.5 and Figure 13.5 from Still Pictures; Figure 1.8 from *Politics of the Earth, Environmental Discourses* © John S. Dryzek 1991, by permission of Oxford University Press (Dryzek, J. S., 1997); Figure 2.2 and Figure 2.4 from TRIP Photo Library; Figure 3.2 from *Modern Geographical Thought*, reproduced with permission of Blackwell Publishers Limited (Peet, R., 1998); Figure 4.2 and Figure 4.3 courtesy of the Oriental Institute, University of Chicago © 2000; Figure 4.4 from the National Arts Slide Library Record © De Montfort University; Figure 4.5 from Sustainable Agriculture in the Middle Ages, the English Manor in *The Agricultural History Review*, Vol. 38 No. 1, 1–19 (Pretty, J., 1990); Figure 4.6 from *The Making of the English Landscape*, reproduced by permission of Hodder and Stoughton Limited (Hoskins, W. G., 1955); Figure 4.7 from *Introduction to World Forestry*, reproduced with permission of Blackwell Publishers Limited (Westoby, J., 1989); Figure 5.1a from *Reconstructing Nature, Alienation, Emancipation and the Division of Labour*, published by Routledge, London (Dickens, P., 1996); Figure 5.1b from *Environmental Problems, Nature, Economy and State* © John Wiley & Sons Limited. Reproduced with permission (Johnston, R. J., 1989); Figure 5.2 from Atmosphere Picture Library; Figure 6.1 from *Australia* (The World's Landscape Series), reprinted by permission of Pearson Education Limited © Longman Group Ltd 1976 (Heathcote, R. L., 1975); Figure 6.3 from *Political Geography, World Economy, Nation-State and Locality* (2nd edn) reprinted by permission of Pearson Education Limited (Taylor, P. J., 1989); Figure 6.4 provided by the National Museum of Labour History, Manchester; Table 7.1 from *A World of Difference: Society, Nature, Development*, reproduced with permission from The Guilford Press (Porter, P. and Sheppard, E., 1998); Figure 7.1, Figure 7.2 and Figure 7.7 from Hulton Getty Picture Collections; Table 7.2 from *The Agricultural Pattern in The*

Changing Geography of the UK, Johnston, R. and Gardner, V. (eds) published by Routledge, London (Bowler, I., 1991); Figure 7.4 from *The Environmental Heritage of Soviet Agriculture*, CAB International, Wallingford (Libert, B., 1995); Figure 7.6 reproduced from the web site of the Space Monitoring Information Support Laboratory, Moscow; Figure 8.1 and Figure 8.2 from *Climate and Human Change: disaster or opportunity*, Parthenon Publishing (Cowie, J., 1998); Figure 8.3 reprinted by permission from Nature (379: 240) © 1996 Macmillan Magazines Ltd; Figure 10.3 reproduced with permission from the web site of the Canadian Forest Service; Figure 11.2 reproduced with permission from *Planning for a Sustainable Environment*, Town and Country Planning Association (Blowers, A., 1993); Figure 11.3 reproduced with permission from *T&CP Special Supplement*, Town and Country Planning Association (Hall, P., 1998); Figure 12.4 reproduced with permission from Action Aid; Figure 12.8 from Adrian Arbib; Figure 13.7 from Hertfordshire County Council; Box 14.1 John Brick cartoons from *What We Consume*, Unit 10, Activity 9.1, published by WWF/Richmond (1990).

Whilst every effort has been made to trace the owners of copyright material, in a few cases this has proved impossible and we take this opportunity to offer our apologies to any copyright holders whose rights we may have unwittingly infringed.

Chapter 1

Ecology, society and environment

1.1 Introduction

Welcome to *Environments in a Changing World*. You may be one of the first-year undergraduate students of geography for whom this book was primarily written, or you may be part of a more general readership with a developed interest in the environment and environmental issues. Either way, we hope to introduce you to the social processes that shape premodern, modern and postmodern environments; the manner in which these environments shape continued social development; and the prospects for creating more healthy and fulfilling environments for greater numbers of the world's people in the future. On the way we will explore a range of environmental issues, trace evolving geographical thought, and encourage you to take an informed and active role in environmental politics.

   At the core of much contemporary concern about the environment is the relationship between society and nature that has long preoccupied geographers. This introductory chapter challenges you to think about that relationship in what may well be new and challenging ways. It clarifies the concepts of environment and nature, and introduces sustainability in the context of ecological limits that afford both challenges and opportunities. By introducing the differing relations between ecological and social processes that characterize premodern, modern and postmodern societies, it also sets the scene for subsequent chapters.

1.2 The environment

Put simply, the **environment** is our surroundings or where we live. It is the changing reality, both material and non-material, that shapes our development and well-being, and is the product of both ecological and social relations and processes. **Ecological relations** are those between living organisms and their non-living environment: between plants, animals, rock, soil, air and water. These relations structure the biophysical world and make possible such processes as photosynthesis, respiration, biological evolution and the functioning of ecological systems. Physical geographers are among the environmental scientists who study ecological relations and systems, and it is likely that your studies of geography to date have included such topics as weather and climate,

1

soil-formation and classification, energy flow and material cycling in eco-systems and natural regions or biomes. You will perhaps be surprised to find little reference to physical geography and the science of ecological relations in this human geography book once the nature of ecological limits have been explored later in this chapter. If you require a short introduction to ecology or wish to explore further the ecological relations shaping the issues dealt with in subsequent chapters, there are many texts that can help you (Goudie and Viles, 1997; Miller, 1995; Odum, 1997).

Environments are also the product of social relations and processes. **Social relations** are a distinct subset of ecological relations made possible by the ability of human animals to organize themselves in ways that sustain their well-being. At different times and in different places, people organize them-selves differently to produce the goods and services they need to survive and develop: to reproduce people; and to reproduce such conditions of production as those resources (e.g. clean water) and services (e.g. waste recycling) pro-vided by ecosystems. People enter willingly or unwillingly into class, gender, political, cultural, spatial and other kinds of social relations, and these make possible such social processes as habits, customs, laws, institutions, language and ideologies. Social relations structure our behaviour in the environment but are also the result of our actions or agency.

Consideration of climate change or genetically modified food reminds us that social and ecological processes affect one another constantly. Society is nature (a subset of ecological relations) and nature is social (ecological relations and processes are almost universally affected by social relations and processes). Like the human body, the environment is a hybrid. It is both natural and social, object and subject (the result and cause of processes), and material and discursive (hard reality and the subject matter of language, text and symbols of all kinds). Consideration of the social construction of nature supports this argument.

1.3 The social construction of nature

For students educated in a culture that separates nature from society, and encourages specialization in either the natural or social sciences, it may be difficult to accept that the environment (everything around us) is nature or natural and is socially constructed or partly the result of social processes. Figure 1.1 represents **the construction of socionature** (socialized nature), and you will see that this is the result of both material and representational prac-tices. Cultural (social) practices work with biochemical/physical processes to produce the objects, subjects and environments that are more or less useful for our survival and development, while at the same time language and discourse is constructed to enable, describe, explain or conceal what is happening. The space we create and reside in (the environment) is the product of such ma-terial and representational practices as working, eating, reproducing, loving, playing, educating, advertising and entertaining, and these are shaped by mul-tiple and often contradictory social relations and ideological practices. Once

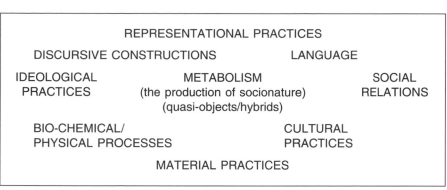

Figure 1.1   The construction of socionature
Source: Swyngedouw, 1996.

created, productive, domestic, recreational, educational and other environments act back on society to regulate the ongoing metabolism of society and nature. Environments help to regulate the interactions between people and between people and the rest of nature. They are barriers or facilitators of different kinds of action or development, and are consequently often both the cause and site of social conflict and struggle.

Swyngedouw (1996) explains the metabolism that lies at the heart of Figure 1.1 and illustrates it by reference to drinking a cup of water from a New York stand-pipe.

> I mean that the 'world' is a historical–geographical process of perpetual metabolism in which 'social' and 'natural' processes combine in an historical–geographical 'production of socionature', whose outcome (historical nature) embodies chemical, physical, social, economic, political and cultural processes in highly contradictory but inseparable manners. Every body and thing is a cyborg, a mediator, part social and part natural, but without discrete boundaries, continually internalising the multiple contradictory relations that redefine and rework every body and thing. Again, my cup of water can serve to exemplify this. Drinking water from the stand-pipe combines the circulation of productive, merchant and financial capital with the production of land rent and their associated class relations; the ecological transformation of hydrological complexes and the bio-chemical process of purification with the libidinous sensation and the physiological necessity of drinking fluids; and social regulation of access to water with images of clarity, cleanliness, health and virginity. Although I cannot separate these 'concepts' and practices from each other in the flow of water, it does not take much to identity the profound social, cultural, political and ecological forces, struggles and power relations at work in this perpetual metabolizing circulation process of flowing water (Swyngedouw, 1996, p. 70).

Notice how water is the product of both ecological and social relations and processes; how the production and consumption of water is both material and representational; how access to water is regulated and celebrated with the aid of images; and how the metabolism involves complex social struggles. You might like to consider the ways in which reading this textbook connects you to a similar web of relations and processes.

Such consideration should convince you that no account of environments in a changing world can be somehow objective, impartial or neutral. All representation of socionature, including geography books, is a product of the metabolism it seeks to describe, explain or reconstruct. We will set out our own ethical, philosophical and political positions later. For the moment we need to return to **the concept of nature**.

1.4 So what is nature?

At this point you are probably thinking 'hold on a minute, surely some environments are untouched by social relations and processes and can be regarded as completely "natural" rather than partly socially constructed?' There are clearly environments, often termed wilderness, where the impact of society is small or minimal, but there is no place on the surface of the Earth where human impact cannot be detected in the soil, air, water or wildlife. All environments and natures are partly the product of social processes and this leads McKibben (1990) to write of the death of 'nature' and Smith, N. (1984; 1996; 1998) to distinguish between first, second and third natures. **First nature** is the given, pristine nature of biophysical processes, laws and forms existing prior to its contact with society. It is the nature from which society evolved, but the impact of society is now so all-pervasive that first nature has everywhere been replaced by second nature. **Second nature** is culture and environment, part biophysical, part social. It takes such forms as the human body, the city, the countryside and genetically modified organisms, and these reveal how second nature continues to be produced out of second nature. **Third nature** is the information flows and representational signs, symbols and discourses of nature that pervade our everyday lives in such forms as advertisements, television documentaries and Internet pages. Such media acknowledge and represent the social construction of second and third nature more or less realistically and truthfully for they continue to be influenced by **ideologies of nature**.

Broadly speaking, an **ideology** is any set of ideas that renders the world more comprehensible, contains taken-for-granted ('common-sense') assumptions that are never questioned, and is 'taught' through such channels as education, the media and everyday lived experience. As Pepper argues:

> More narrowly defined, the term denotes a set of ideas, beliefs and ideals, which are the basis of an economic or political theory or system. If the assumptions are brought to the surface they are generally seen to reflect the material vested interests of those who share the ideology (for instance the assumption that competition is natural often underpins the ideology of those who do well in economic competition). Ideologies are presented as statements of universal truth, but reflect these narrower interests (Pepper, 1996, 331).

By suggesting that there continues to be a first nature, separate from society, such interests can attribute to nature characteristics and processes that are partly or mainly social. They can blame nature for poverty, disasters, or the low status of women, and can legitimate oppressive social relations as being somehow natural or reflecting natural values. By failing to acknowledge that human and other forms of nature are socially constructed, ideologies of nature can encourage fatalism and an acceptance of the status quo. We will return to such ideologies later in the chapter when we outline the discourses shaping contemporary environmental politics.

Having outlined the social construction of nature and warned against belief in a first nature outside society, how can we best understand the **multiple meanings of nature** found in everyday and academic language? Soper (1995) provides an answer by distinguishing between three concepts:

- **The metaphysical concept of nature** whereby humanity thinks its difference and specificity. The concept of the non-human (even though the humanity–nature demarcation has been disputed and our ideas of what is 'nature' are continually revised as we change our perceptions of what is 'human'). The nature which is opposed to the 'human' or 'cultural'; the nature that is invoked whenever we pose the question of humanity's relation to nature.
- **The realist concept of nature** whereby 'nature' refers to the structures, processes and causal powers that are constantly operative within the physical world, that provide the objects of study of the natural sciences and condition the possible forms of human intervention in biology or interaction with the environment. The nature to whose laws we are always subject, even as we harness them to human purposes, and whose processes we can neither escape nor destroy.
- **The lay or surface concept of nature** used in much everyday, literary and theoretical discourse. 'Nature' refers to the ordinary observable features of the world: the 'natural' as opposed to the urban or industrial environment; animals domestic and wild; the physical body in space; and raw materials. The nature of immediate experience and aesthetic appreciation; the nature we have destroyed and polluted and are asked to conserve and preserve.

Nature is used to help us think our difference from the rest of the world; to refer to those ecological relations, processes and limits that inevitably shape our lives; and to defend an environment that sustains the culture of everyday life. Gorz employs the lay or surface concept of nature when he argues that much environmental politics is a protest against the destruction of the lifeworld by the workings of the formal economy and the state. The defence of the lifeworld, the culture of everyday life or civil society, entails defending environments that sustain knowledge, habits, norms and modes of conduct that enable individuals to interpret, understand and assume responsibility for the way in which they inhabit the world that surrounds them. Such environments appear 'natural' 'because their structures and workings are accessible to intuitive understanding; because they correspond to a need for a flowering of sensory and motor faculties; and because their familiar structure enables individuals to find their way about in them, interact with them, and communicate with them "spontaneously" using aptitudes that have never been formally taught' (Gorz, 1993, pp. 57–8).

Our consideration of ecological and social relations has already hinted that this text will be strongly guided by the realist concept of nature. Such nature does limit social development, as further consideration of the metabolism of socionature reveals.

1.5 Entropy and ecological limits

All human activity depends on energy and material flows within ecological systems. Ecosystems are powered by sunlight, build living material as energy moves through food-chains, and recycle their wastes. They can provide economic systems with a renewable supply of low-entropy matter-energy, in such forms as timber, draught animals or fish-stocks, and past environmental conditions have resulted in some of this being stored in such non-renewable forms as coal or oil. Ecosystems also provide services by, for example, cleaning air and water, stabilizing soils, regulating climate, controlling pests, recycling wastes, and acting as repositories of genetic information. The true value of these resources and services, or **ecological capital**, is rarely acknowledged in economic accounting.

The **low-entropy matter-energy** produced by ecosystems is all ultimately derived from sunlight. It is capable of doing work: but the second law of thermodynamics (a feature of realist nature) means that whenever it is converted into work, its entropy increases or the availability of usable energy declines (Rifkin and Howard, 1985; Daly and Cobb, 1990). Burning coal or food releases dispersed forms of energy or heat but the energy required to reconcentrate this energy is greater than that which could be recovered. The analogy of sand in an hour-glass may help you to understand the limits imposed on society by the second law. Our metabolism with nature is dependent upon low-entropy matter-energy (sand in the top of the hour-glass) and once this has been used up (the sand has run through to the bottom) the energy required to turn the hour-glass upside down, or convert higher-entropy matter-energy back into lower-entropy energy-matter, is greater than the energy it would make available. The economic processes that lie at the heart of the construction of socionature depend on continuing inputs of low-entropy resources and produce continuous outputs of high-entropy waste. We live on the qualitative difference between resources and waste (the increase in entropy) and our social systems are more or less efficient at using technology to convert this entropy gain into wealth and well-being.

The are two sources of low-entropy energy-material: the solar and terrestrial. While the solar is practically unlimited in stock but strictly limited in its flow-rate of arrival to Earth, the terrestrial is strictly limited in stock but society can use **technology** (tools of all kinds including language and ideas) to increase the flow-rate. Premodern economies operated within or alongside the cycles of nature. They depended on the long-lasting supply of solar energy captured and made available by ecosystems, used mainly renewable materials, and generated little waste. Figure 1.2 suggests that their reach was local or regional but that premodern agricultural societies did overload some regional ecosystems by over-exploiting their resources or carrying capacity. The economic surplus

Era	Resources		Sinks	Productivity and surplus	Value/commodity/money	Societal reach *Geographical range*	Social type of society	*Chapters in this text*
	Matter	Energy						
Premodern pre-neolithic	renewable	solar	no overload of carrying capacity of local ecosystems	no or small surplus	rudimentary/ reciprocity	local	hunters and gatherers	*3, 4*
Premodern neolithic	Renewable/ non-renewable	solar	partial overload of regional ecosystems	increase of productivity; surplus production in limits of agriculture	partial monetization/ reciprocity and equal exchange	regional	agriculture	*3, 4*
Modern industrial	Renewable/ non-renewable	fossil	heavy overload of global ecosystems	social rule of permanent increase of productivity; surplus in form of surplus value	full monetization; reciprocity as a complementary rule	global	industry	*5, 6, 7, 8, 9, 10, 11, 12*
Postmodern post-industrial	renewable	solar	reduction of load to limits of carrying capacity of global ecosystems	ecological productivity; surplus in socialized form	partial monetization; rule of reciprocity	global	post-industrial society	*9, 10, 11, 12, 13*

Figure 1.2 Some characteristics of premodern, modern and postmodern societies

Source: Altvater, 1998, p. 26. (Altvater entitled this table 'Brief characteristics of Promethean revolutions in the history of mankind'. The text in italics has been added.)

that they produced facilitated the replacement of barter or reciprocal exchange by money, market exchange and an eventual industrial revolution.

Modern industrial societies operate linear economies, somewhat apart from the cycles of nature. They depend on fossil energy, use much non-renewable material, and generate considerable waste. **Industrialism** commits such societies to continuous growth of productivity, economic output and surplus in the form of wealth or profit. Consequently they use technology to overcome the constraints of space, time and ecology. Their reach becomes global, economic processes are speeded up, and the use of ecological resources and sinks becomes increasingly unsustainable. Significant social change, accelerating over the last 30 years, marks the arrival of postmodern or post-industrial societies with information-based technologies and economies. These provide new opportunities to restore our dependence on sunlight and renewable materials, and reduce our demands on ecological services and sinks to levels well within the carrying capacity of global ecosystems. The postmodern challenge is to create sustainable natures in new and liberating ways, and as Figure 1.2 suggests this may entail a greater degree of socialization or common ownership of economic surplus. Box 1.1 illustrates the changing constructions of nature, outlined in Figure 1.2, through a case-study of Bedfordshire's Marston Vale.

Box 1.1   The Marston Vale

The Marston Vale, Figure 1.3, lies between Bedford and Milton Keynes and is crossed by the London to Birmingham motorway (M1) and the London to Sheffield railway (Midland Mainline). The Vale is formed on Oxford Clay, a Jurassic deposit, which runs across England from Dorset in the southwest to the Yorkshire coast in the northeast. The exposures of Lower Oxford Clay in the Vale and adjoining areas of Buckinghamshire and Cambridgeshire are particularly suitable for brick manufacture.

The Vale is very flat with an average elevation of around 50 m. It is drained by the Elstow Brook into the River Great Ouse to the north, and bounded to the south by the steep scarp slope of the Lower Greensand. Soils are heavy clay loams that are difficult to cultivate, and prior to human settlement much of the Vale was covered by oak and ash forest.

Premodern manors and feudal villages

By the time of the Domesday Survey in 1086 much forest had been cleared to create a number of largely self-sufficient villages, each with a mix of meadow, marsh, arable, waste and woodland. Pressure from increasing numbers of people was accelerating clearance, but forest remained a key resource providing building materials, fodder for livestock, herbal remedies and fuel. The survey was carried out so that William the Conqueror could tax people according to the carrying capacity of their land (woodland was valued according to the number of swine that it could support). In feudal society, lords, villeins, bordars, serfs and socmen had different rights and obligations to one another and the land.

(continued)

(continued)

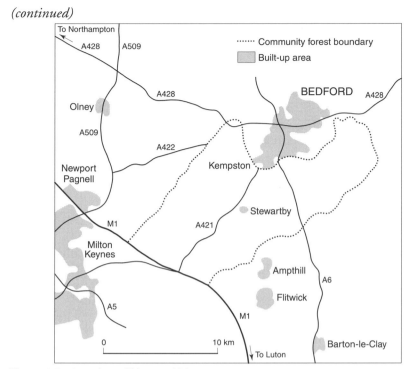

Figure 1.3 Location of Marston Vale
Source: Bedfordshire Country Council.

The gradual intake of waste and woodland to manorial estates, together with harsh forest laws, caused much resentment among the poor.

Modern brickfield

In the 1880s a revolution in brick-making took place in the village of Fletton near Peterborough. It was discovered that Oxford Clay was ideal for moulding and firing into bricks. It contains oily combustible material and virtually fires itself. Brickworks, based on the Fletton process, spread to the Marston Vale in the 1890s and rapid growth followed after several firms amalgamated with the London Brick Company in 1923. Large works supplying a regional market replaced small works supplying local markets, and there were attendant problems of air pollution and land degradation.

The Stewart family, local owners, built the model village of Stewartby for brick-workers in 1937. Production in the Vale reached a peak in 1954 (two billion bricks) and throughout the 1950s and 1960s, labour shortages were met by migrant workers from Europe, the Caribbean and Asia. With changes in building technology the demand for bricks declined from the late 1960s, and by 1974 there were only three works still operating. The Vale had become a 'problem environment' characterized by air pollution and huge pits from which clay had been extracted (Blowers, 1984). These were increasingly used

(continued)

(continued)

for 'landfill', the disposal of domestic and other waste. In 1984 the London Brick Company became part of the Hanson Trust.

Postmodern community forest

In 1990 the government introduced the community forest concept and designated 12 such forests, including the Marston Vale. The community forest brings together private business and farming interests, local communities and local councils, to tackle planning and environmental problems in a new spirit of partnership. Government grants, including money from the national lottery and the landfill tax, are matched with money from the County Council, local landowners and business.

Covering 16,000 hectares the community forest will increase tree cover in the Vale to around 30 per cent. The forest team are encouraging more sustainable forms of development with new opportunities for forestry and farming, recreation, education and new habitats for wildlife. A visitor centre, Figure 1.4, trails and other facilities have been developed. The forest provides farmers with new forms of subsidy and Shanks, the major landfill operator, with new opportunities to 'green' its image. This company generates power from landfill gas schemes, which have attracted grants from the Department of Energy. Many of those who live in the Vale now commute long distances to work, but there are new enterprises nearby, including the national distribution centre of the Internet bookseller Amazon. The expected expansion of Milton Keynes will bring the new city to the western margins of the forest sometime in the twenty-first century.

Figure 1.4   Marston Vale Forest Centre
Source: John Huckle.

1.6 The four-capital model of environmental economics _____

The nature of our current predicament, and the challenges and opportunities it presents, can be further understood by reference to the model of wealth-creation shown as Figure 1.5 (Ekins, Hillman and Hutchison, 1992). While conventional economics recognizes three forms of capital (land, labour and such manufactured forms of capital as factories and machines), this model replaces land with ecological capital and adds social and organizational capital. **Capital** is the broad stock of resources of all kinds that makes possible and generates a flow of production (goods and services). Figure 1.5 shows how some of this flow has to be reinvested to replace capital that is worn out or used up. The rest can be regarded as surplus or income, and can be consumed. Whereas conventional economics measures people's utility or welfare by their level of consumption or income (e.g. Gross National Product (GNP) per capita), the four-capital model shows welfare to be derived from many sources. It also suggests that income and consumption can only be increased at the expense of reinvestment and that over-consumption will eventually lead to long-term deterioration of capital stocks. Wealth can only be produced on a long-lasting or sustainable basis if all four capitals are conserved, including that stock of low-entropy matter-energy associated with ecological resources and services.

You should spend some time studying Figure 1.5, analysing the multiple flows that contribute to human welfare and considering the implications of failing to invest adequately in the maintenance of each kind of capital (Figure 1.6). You should then consider the facts and figures in Box 1.2. which give some indication of the state of the four kinds of capital in the contemporary world. Clearly modern forms of development that colonized the world in recent centuries, are not sustainable. While they have undoubtedly improved the welfare of many people and created huge stocks of manufactured capital, they have also resulted in a global society in which the rich live at the expense of the poor (much human capital remains underdeveloped), community self-help and governance is undermined (social and organizational capital is eroded), and ecological resources and services are over-exploited (ecological capital is damaged or destroyed).

Contemporary global society is unsustainable because the ecological relations required to provide for ever higher levels of consumption and manufactured capital are destroying species and their habitats. The destruction of ecological capital threatens the continuing supply of materials that feed the production process and the conditions, such as fertile soil, uncongested space and healthy workers, that make production possible. Ecological problems contribute to economic problems, and economic and political elites may then further reduce the flow of wealth or welfare going to the poor. Social relations show growing divisions between rich and poor at all scales resulting in real and potential social conflict. Ecological, economic and social unsustainability are compounded by cultural unsustainability as local livelihoods and knowledge are destroyed, and personal unsustainability as people experience physical and mental ill-health.

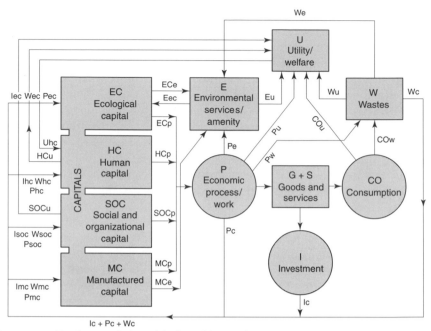

Figure 1.5 The four-capital model of wealth-creation
Source: Ekins, *et al.*, 1992.

Figure 1.6 A natural disaster? Human capital suffers the effects of degrading ecological capital. Orphaned survivors of flash floods in the Phillipines
Source: Still Pictures.

Key to Figure 1.5

Key

EC	**Ecological capital** The biophysical resources and services that contribute to human welfare
ECe	Ecological services provided independent of human activity e.g. climate regulation, aesthetically pleasing landscapes
Eec	The impact of environmental services on renewable resources e.g. desertification
Ecp	Resources for production, renewable (e.g. timber) and non-renewable (e.g. oil)
Iec	Investment in ecological capital e.g. nature reserves
Wec	The environment absorbs, neutralizes, or recycles much waste from the economy
Pec	The impacts of the economic process on ecological capital
HC	**Human capital** The health, knowledge and skills that enable people to contribute to economic production and social reproduction
HCp	The productive work that humans do in the economy
Uhc	An unhealthy, unskilled and unmotivated person lacks welfare or utility
Hcu	A healthy, skilled and motivated person enjoys satisfaction, welfare or utility
Ihc	Investment in human capital e.g. education and training
Whc	The impact of wastes on human capital e.g. pollution and health
Phc	The impacts of economic process on human capital
SOC	**Social and organizational capital** The characteristics of social organizations that enhance the productivity of human capital
SOCp	Social and organizational capital that contributes to the economic process
SOCu	Social relations embodied in social and organizational capital at home, in the workplace and community contribute to human satisfaction and welfare. People's organizations use social capital to develop powers to defend their rights and improve the quality of life
Isoc	Investment in social capital e.g. by employing management consultants
Wsoc	The impacts of wastes on social capital e.g. local campaigns against pollution
Psoc	The impacts of the economic process on social and organizational capital
MC	**Manufactured capital** All the tools, machines, buildings, technologies and infrastructure that enhance productivity (what conventional economists refer to as 'capital')
MCp	The contribution of manufactured capital to the economic process
MCe	Manufactured capital's contribution to environmental services and amenity
Imc	Investment in manufactured capital
Wmc	The impacts of wastes on manufactured capital
Pmc	Depreciation of manufactured capital through wear and tear
E	**Environmental services/amenity**
Eu	Environmental services that contribute to the satisfaction of human needs
W	**Wastes**
Pw	Wastes which derive from production
Cow	Wastes which derive from consumption
Wc	The impacts of waste on capital of all kinds
Wu	The direct impact of waste on human welfare e.g. litter
We	The impact of waste on environmental services/amenity e.g. global warming
P	**Production**
Pe	Effects of production that detract from environmental amenity e.g. noise pollution
Pu	The contribution of work to human welfare via job satisfaction and working relationships
Pw	Wastes which derive from production
Pc	The impacts of the economic process on capital of all kinds
I	**Investment**
Ic	Investment in capitals of all kinds

Box 1.2   The state of the world

Ecological capital

There are between three and thirty million species on the planet – nobody really knows. However, species are being extinguished by humans up to a thousand times the 'normal' rate and the conservative guess is that one-quarter of all the world's creatures could perish in the next 100 years. All human economy rests on wildlife, and if industry had to write cheques for recycling dead things into nutrients, for cleansing water and converting foul air into clean air, for the prevention of erosion and flood and provision of raw materials, the annual bill would top $30 million million. Scientists at Cambridge University argue that it would cost a mere $320 billion a year to protect biodiversity. The bill is quite small compared with the $1,000 billion subsidies for fishing, road use, agriculture and energy consumption, all of which actually threaten wildlife.

The Living Planet Report from the World Wide Fund for Nature (Loh *et al.*, 1998) suggests that humans have destroyed more than 30 per cent of the world's natural wealth since 1970. As global consumption pressure on ecosystems doubled between 1970 and 1995: freshwater ecosystems declined by 50 per cent, marine ecosystems deteriorated by 30 per cent, and the world's natural forest cover declined by about 10 per cent. Degradation of ecosystems equates to an economic loss of about $1 million million per year. Today humans take for their use somewhere between one-quarter and one-half of all plant material that grows on Earth each year. Fully half of all the atoms of nitrogen and phosphorous that are annually fixed in new plants come from human intervention in the form of fertilizers rather than natural cycles. Global food production doubled over the past 35 years; there is 25 per cent more food per person today than 35 years ago. In the sea we take 10 per cent of all its annual production, and more like 30 per cent in rich areas of nutrient upwelling. Over the next 25 years the world's people will begin to run out of fresh water.

Human capital

In the 1950s average life-expectancy at birth, around the globe, was 46 years. Today it is 64 years. Over the same period the average difference in longevity between the developed and developing world shrank from 26 to 12 years. Partly as a result of such global changes in average health, human numbers have shown unprecedented growth. The world population of six billion in 2000 represents a 60 per cent increase over just the past 35 years. Almost half the planet's new residents live in cities.

(continued)

(continued)

We live in an unequal world. Twenty per cent of the global population accounts for 86 per cent of global consumption, and one billion people have been left out of the consumption boom of the past two decades. Consumption has increased sixfold in the last 20 years and doubled in the last ten. People in Europe and North America now spend $37 billion a year on pet food, perfumes and cosmetics: enough to provide basic education, water and sanitation, basic health and nutrition for all those now deprived of these needs and still leave $9 billion over. The 200 richest people in the world have a combined wealth of more than $1 trillion, equal to the annual income of the poorest 47 per cent of the Earth's population: some 2.5 billion people. Among the 4.4 billion people in developing countries: almost three in every five lack basic sanitation, one-third have no safe drinking water, a quarter have inadequate housing, while one-fifth are undernourished. One-fifth of the world's population lives on less than $1 a day.

Social and organization capital

There has been an explosion in recent decades of small and large pressure groups, advisory agencies, research groups, aid charities, 'network' associations, professional bodies . . . thousands of non-governmental organizations (NGOs), many with memberships and budgets spanning several countries. Like workers' organizations, these citizens' organizations seek to influence the policies of private corporations, governments and international agencies. They contribute to the governance of societies at all levels, and some form a global coalition supporting more sustainable forms of development.

The United Nations Development Programme (UNDP) is upbeat about the improved human rights climate in many countries and is promoting more direct forms of democracy. While more countries are ratifying the various United Nations (UN) conventions on human rights, there has been a decline in the 'quality of governance' in many countries. More than 200 of the 850 experts consulted by the United Nations Environment Programme, when compiling its *GEO 2000* report (UNEP, 1999), mentioned poor governance as an emerging issue for the twenty-first century. Many environmental treaties, such as those on substances that damage the ozone layer, acid rain in Europe and the export of toxic waste, have been successful, but the report suggests the UN remains 'too weak' in a lot of areas.

Many NGOs, together with labour unions, students, anarchists and others, form the global anti-corporate movement – an amalgam of environmentalism, anti-capitalism, anarchy and much else. Its activism is directed at large corporations, governments and international agencies; seeks direct democratic or community-based decision-making; and mirrors the organic, decentralized and interlinked pathways of the Internet. The movement opposes the global

(continued)

(continued)

deregulation that concentrates power in fewer hands, prefers a 'surfer's approach' to politics to an overarching revolutionary philosophy and is skilled in its use of media and information technology.

Manufactured capital

Between 1950 and 1990 the real output from the global economy increased fourfold. In the last 40 years marine fish consumption has more than doubled; wood and paper consumption has increased by two-thirds; cement production has increased more than fourfold, and carbon dioxide emissions have doubled.

If the world economy grows by 2–3 per cent per annum, which implies a quadrupling of consumption by 2050, and world population doubles over the same period, then the environmental impact per unit of consumption (a measure of how efficiently the economy uses natural resources and produces wastes) will have to be one-sixteenth of its current level by 2050 to realize significant reductions in the economy's environmental impact. In other words, technologies and living patterns will have to be 91 per cent more environmentally efficient than they are now. This is essentially a challenge to the rich countries of the North since the extra projected 60 million Northerners will be responsible for more environmental degradation that the 900 million additional people who will live in poorer countries. Since living standards must rise in the South, the improvement in environmental efficiency in the North will probably have to be greater than 91 per cent.

Sources: Jacobs, 1996; Klein, 2000; Loh *et al.*, 1998; UNDP 1998; UNEP, 1999.

Before considering the nature of more sustainable forms of development, it is necessary to note that conventional economics assumes that capitals can be easily substituted for each other and that the market will take care of any problems created by the decline of ecological capital. It does not matter if such capital is destroyed so long as manufactured capital is built up (e.g. a housing estate with a landscaped lake replaces a wetland area). Rising prices, together with consumer demand, will prompt technological innovation to find equivalent resources and services (e.g. fish can be farmed, crops can be genetically engineered, water can be artificially cleaned). Such arguments neglect the importance of **critical ecological capital** (e.g. biodiversity) that is irreplaceable and should be preserved; fail to take account of the second law of thermodynamics; and fail to acknowledge limited **substitutability** between flow and fund resources (Georgescu-Roegen, 1971). While there may be substitutability between the capital and labour (fund resources), or among various flows of low-entropy resources from ecological systems, there can be very little substitutability between funds and flows since these are complements, not substitutes, in the process of production. Daly and Cobb (1990) give the example of house-building. A house can be built with fewer carpenters and more power-saws, but no amount of carpenters or power-saws will allow a significant reduction in the amount of timber or nails. A sustainable society must live within ecological limits shaped by the availability of ecological capital and the flow of low-entropy matter-energy that this can provide. In such a society the rate of entropy-production should be zero: energy inputs from the Sun should balance out the increase in entropy in such forms as heat, sewage and waste discharges.

1.7 Sustainable development

Something that is sustainable can last indefinitely. Sustaining the metabolism between ecological and social systems (Figure 1.1) or sustaining systems of wealth-production (Figure 1.5) means sustaining three sets of relations (Hartmann, 1998):

1. Relations among humans based on mutual respect and tolerance. Just relations allow equitable access to food, clothing, health care and meaningful work, provide for freedom of thought and mental development, and promote democratically determined economic and political decisions.
2. Relations among humans and other species that minimize human impact on other species and their environments or habitats.
3. Relations between organisms and their environment that have created the climate, water cycle, radioactive levels and other environmental conditions (ecological processes) that we have experienced throughout most of human history.

Creating and maintaining these relations requires us to care for the welfare of other human beings, future generations and other species, and requires us to translate this concern into appropriate forms of governance and citizenship. By translating appropriate environmental ethics into effective laws and institutions (Alder and Wilkinson, 1999), we could prevent the destabilization of climate and the ozone layer; conserve critical ecosystems and biodiversity; ensure sustainable harvesting of renewable resources and soil and water conservation; use non-renewable resources intensively via durable design, repair, reuse and recycling; match the level of use of non-renewables to the level of discovery via depletion taxes; keep emissions within the capability of the environment to absorb, neutralize and recycle; and minimize risk via the precautionary principle that may require us to forgo dangerous technologies.

Advocacy of **sustainable development** has grown rapidly since the late 1980s, and Figure 1.7 provides some definitions and indicators. Such development seeks to reconcile economic development with environmental protection, but there is no consensus on the ethics or politics of sustainability (Reid, 1995).

Sustainable development is development that meets the needs of the present without compromising the ability of future generations to meet their own needs (WCED, 1987).

Sustainable development means improving the quality of life while living within the carrying capacity of the supporting ecosystems (IUCN/UNEP/WWF, 1991).

The core meanings of the term now have wide agreement. They are that the environment must be protected in such a way as to preserve essential ecosystem functions and to provide for the well-being of future generations; that environmental and economic policy must be integrated; and that the goal of policy should be an improvement in the overall quality of life, not just income growth; that poverty must be ended and resources distributed more equally; and that all sections of society must be involved in decision-making (Jacobs, 1996, p. 26).

General indicators of a sustainable community:

1. Resources are used efficiently and waste is minimized by closing cycles.
2. Pollution is limited to levels which natural ecosystems can cope with and without damage.
3. The diversity of nature is valued and protected.
4. Where possible local needs are met locally.
5. Everyone has access to good food, water, shelter and fuel at reasonable cost.
6. Everyone has the opportunity to undertake satisfying work in a diverse economy. The value of unpaid work is recognized, while payments for work are fair and fairly distributed.
7. People's good health is protected by creating safe, clean, pleasant environments and health services which emphasize prevention of illness as well as proper care for the sick.
8. Access to facilities, services, goods and other people is not achieved at the expense of the environment or limited to those with cars.
9. People live without fear of personal violence from crime or persecution because of their personal beliefs, race, gender or sexuality.
10. Everyone has access to the skills, knowledge and information needed to enable them to play a full part in society.
11. All sections of the community are empowered to participate in decision-making.
12. Opportunities for culture, leisure and recreation are readily available to all.
13. Places, spaces and objects combine meaning and beauty with utility. Settlements are 'human' in scale and form. Diversity and local distinctiveness are valued and protected (LGMB, 1995).

Figure 1.7   Sustainable development

1.8 Environmental ethics and politics _____

People are constituted by and constitute both ecological and social relations. They are part of ecological relations (members of a biological species, dependent on ecological resources and services to supply their needs) yet partly independent of such relations as part of social relations (they have powers of language and technology that enable them to transform their inherited natures and the nature that surrounds them). Since they can apply their growing understanding of the world to the regulation of ecological and social relations, by means of appropriate technology and democratic forms of global governance and citizenship, there are grounds for cautious optimism concerning progress towards more sustainable futures.

When considering the ethics that should guide such progress, we should note that our position within ecological and social relations gives rise to a tension, seen in art, literature and culture more generally, between **immanence** (the pull of 'nature' or the desire to live according to 'nature') and **transcendence** (the pull of culture or the desire to rise above the harsh realities of 'nature') (Soper, 1999). As a 'mode of thought' **ecocentrism** (O'Riordan, 1981) reflects immanence by basing ethics and politics on a friendly 'nature' that may be assumed to have value in its own right (intrinsic value). Ecocentrics regard humans as just another component of natural ecosystems, subject to ecological and systems laws; respect nature in its own right and for pragmatic reasons; lack faith in the technology and bureaucracy associated with modern societies; and advocate various kinds of self-sufficiency, decentralization, low-impact technology and spirituality.

Technocentrism on the other hand reflects transcendence by basing ethics and politics on the virtues of exploiting 'nature' as a resource. Technocentrics are committed to industrialism and economic growth, and while they recognize environmental problems, they believe that society will always solve them through careful economic and technical management. They have much faith in expert scientists, technologists, economists and environmental managers, and show little desire for more radical forms of democracy. Both technocentrism and ecocentrism reflect ideologies of nature (see Chapter 2 on positivism and romanticism) by assuming that there is an autonomous 'nature' (first nature) outside society. But as we have seen, nature is best viewed not as a resource for our use (technocentric materialism), or a source of intrinsic value (ecocentric idealism), but as a social category that is the product of both ecological and social relations (historical materialism) (Martell, 1994).

The philosophical and political challenge is not to choose between immanence and transcendence but to balance the two in a political theory that promotes sustainability and the welfare of both human and non-human nature. In arriving at such a theory, Dryzek (1997) reviews nine **environmental discourses**, that depart from industrialism in more or less radical and imaginative ways, and provide the vocabularies and conceptual frameworks through which environmental concerns are expressed in Western liberal democracies. He introduces these discursive constructions in a way that suggests they are shaped by ideological practices as shown in Figure 1.1:

> A **discourse** is a shared way of apprehending the world. Embedded in language, it enables those who subscribe to it to interpret bits of information and put them together into coherent stories or accounts. Each discourse rests on assumptions, judgements, and contentions that provide the basic terms for analysis, debates, agreements and disagreements, in the environmental area no less than elsewhere. Indeed if such shared terms did not exist, it would be hard to imagine problem solving in this area at all, as we would have to return to first principles continually (Dryzek, 1997, 8).

As a result of his review Dryzek concludes that a blend of democratic pragmatism, sustainable development, ecological modernization and green rationalism provides the most appropriate foundations for an appropriate political theory. Figure 1.8 lists the key assumptions of these discourses and you will find them,

	Democratic pragmatism (leave it to the people)	Sustainable development (environmentally benign growth)	Ecological modernization (industrial society and beyond)	Green rationalism (save the world through new politics)
1 Basic entities recognized or constructed	• liberal capitalism • liberal democracy • citizens	• nested social and ecological systems • capitalist economy • (no limits)	• complex systems • nature as waste treatment • capitalist economy • the state	• global limits • nature as complex ecosystems • rational humans • social, economic and political structures
2 Assumptions about natural relations	• equality among citizens • interactive political relationships, mixing competition and cooperation	• subordination of nature • economic growth, environmental protection, distributive justice and long-term sustainability go together	• partnership encompassing government, business, environmentalists, scientists • subordination of nature • environmental protection and economic prosperity go together	• equality among people • complex interconnections between humans and nature
3 Agents and their motives	• many different agents • motivation a mix of material self-interest and multiple conceptions of public interest	• many agents at different levels, notably transnational and local rather than the state; motivated by the public good	• partnerships motivated by the public good	• many individual and collective actors, multidimensional motivation • agency in nature downplayed though not necessarily denied
4 Key metaphors and rhetorical devices	• public policy as a resultant of forces • policy like scientific experimentation • thermostat	• organic growth • connection to progress • reassurance	• tidy household • connection to progress • reassurance	• organic metaphors • appeals to reason, and potential rationality of social structures • link to progress
5 Reason the discourse is of value in promoting ecological democracy	allows expression of the community interest	allows a plurality of approaches to sustainable development at different scales	can take strong forms that promote institutional change and social learning	radicalizes the other three discourses and anchors them in a utopian realism

Figure 1.8 Four environmental discourses that can contribute to an ecological democracy
Source: Dryzek, 1997.

particularly green rationalism, well represented in this book. The discourses that we, with Dryzek, find less helpful are those that can be summarized as survivalism (looming tragedy), the promethean response (growth forever), administrative rationalism (leave it to the experts), economic rationalism (leave it to the market) and green romanticism (save the world through new consciousness).

Jacobs (1997) explains that while the environment has been accepted by politicians and the public as a key area of concern, alongside the economy, health, education and crime, it differs from these other policy areas in important ways. The environment is not one issue but many, encompassing a vast range of subject matter, interest groups and institutions. They have become grouped together because the environment is the only issue that enters the political arena bearing its own ideology and discourse (Dobson, 1995; Pepper, 1996; Smith, 1999) and supported by its own social movement (**environmentalism**). This movement within civil society has been so successful that governments and business have had to adopt much of its analysis and language, including that of sustainable development. This is both an advantage and disadvantage, for token acceptance has not resulted in the environment becoming central to mainstream politics. There is much agreement between the major parties, but little real argument or radical policy innovation.

1.9 Sustainability and democracy

In reality, the transition to sustainable development involves hard political choices, for ecological and economic sustainability can be realized with more or less social, cultural and personal sustainability. If the former are to be realized with improvements in social justice, cultural diversity and individual fulfilment, social relations in all spheres of people's lives, at all levels, will need to be more democratic. Then people will be able to participate more or less equally in the decisions that affect their lives and will be more likely to realize their common interest in sustainability. An **ecological democracy** (Dryzek, 1997; Mason, 1999) is a democracy without boundaries: across social and ecological systems, across existing governmental jurisdictions, and sometimes requiring new forms of governance constituted as oppositional democratic spheres to existing institutions. Workers' and citizens' movements, coordinated by non-governmental organizations, continue to prompt and pioneer such new forms of global governance and citizenship but their efforts are frustrated by prevailing form of democracy and governance.

While some celebrate the spread of representative or liberal democracy to more nation-states in recent years, it continues to leave major decisions to the market, an elite political class, or a growing number of international institutions without a popular mandate. With increased globalization of economic relations, power has become more centralized and the political spectrum has narrowed as political parties have converged on the centre ground (Swift, 2000). Politicians distrust people's capacity to participate in their own self-government and consumerism encourages passive and uncritical choice. Consequently there is much popular cynicism regarding politicians and politics. The membership of political parties is falling, fewer people are voting, and young people in particular view politicians with apathy and distrust.

In the last two chapters of this book we will link ecological democracy and sustainability to new forms of post-industrial socialism. These would revive and extend democracy by using new technologies and forms of welfare to free people from the demands of work in the formal economy and provide them with time for personal and community development. Before we consider such utopian yet realist scenarios, we have much ground to cover.

1.10 Further reading

The documentary comic-book *Introducing Environmental Politics* (Croall and Rankin, 2000) provides an introduction to many of the ideas in this and subsequent chapters. How does the book's approach to ecology and society differ from that reflected in the geography textbooks you have used to date?

In its attention to the four-capital model used by some environmental economists, the chapter hinted at the differences between conventional and environmental economics. By reference to such sources as Ekins, Hillman and Hutchison (1992), contrast the different approaches taken by conventional and environmental economists to measures of economic development.

Read one or more of the articles in the section on geography and the politics of nature in Barnes and Gregory (1997). These take up ideas introduced in this chapter and explored further in Chapter 2. Pay particular attention to what the authors have to say about nature having some form of resistance (taking its revenge on society), ecofeminism, social constructivism and cyborg life.

1.10.1 Websites

WWF's Living Planet Report: http://www.panda.org/livingplanet/
UN Human Development Report: http://www.undp.org/undp/hdro
UNEP Global 2000 Report: http://grid2.cr.usgs.gov/
Environment links: http://geog-main.lancs.ac.uk/links/ENVIRO.HTM

Chapter 2

Philosophy, geography and environmental knowledge

2.1 Introduction

Before starting on our survey of premodern, modern and postmodern environments and environmental issues, it is necessary to give some attention to the changing philosophies of the natural and social sciences that play a key role in the representation of nature and environment and shape the changing purpose, content and methodologies of geography. This chapter introduces the range of philosophical frameworks which geographers use to research, write and teach about the relations between nature and society, and by advocating an approach based in dialectical materialism, critical realism and critical theory, it further develops the arguments introduced in Chapter 1.

Philosophy is 'the study of the most general and abstract features of the world and categories with which we think: mind, matter, reason, proof, truth, etc. In philosophy the concepts with which we approach the world themselves become the topic of enquiry' (Blackburn, 1996, p. 286). Such concepts shape, and are shaped by, the material and representational construction of socionature (p. 2), and are embedded in the language, discourse and ideology we consciously or subconsciously use to account for the nature and workings of the world and our place and destiny within it. The philosophy of geography studies the concepts that structure geographical knowledge in an attempt to lay bare the subject's foundations and presuppositions. It reveals a diversity of approaches that this chapter will explore.

2.2 Philosophy for geographers

The first point to make about philosophy is that it matters. There is little chance of geographers helping to solve environmental problems and bring about a transition to more sustainable forms of development if they are working with flawed notions of how the world works and how it might be changed. Debates on philosophy in geography revolve around such questions of utility and issues of moral and political purpose. Is the geography I am learning, researching, writing or teaching really useful? Does it reflect such values as rationality and social justice? Does it promote relations between people that encourage them to use natural resources and create environments in more sustainable ways?

	Empiricism	Positivism	Interpretivism	Structuralism	Critical theory	Critical realism
Epistemology How reality can be known. The criteria for judging the truth of a statement about reality.	Knowledge is based in experience.	Knowledge is based in experience supported by verifiable evidence.	Knowledge is created subjectively in a world of meanings created by individuals.	Knowledge is based in the world of structures, processes and relations. Experiences do not necessarily reveal this world.	Knowledge is socially constructed in ways that reflect different interests. The dominance of the technical interest limits understanding.	Knowledge is created by building models of how real processes shape events and experiences in the light of contingent circumstances.
Ontology What it is possible to know. The reality that exists.	The things we experience are the things that exist.	What exists are experiences supported by well established regularities and connections.	What exists is that which people perceive to exist.	What really exists are the structures, processes and relations that shape the world. These cannot be observed directly.	What exists is the possibility of understanding the world through communicative rationality based on consensus.	What exists are the related domains or levels of real processes, actual events, and empirical experience.
Methodology A set of rules and procedures to guide enquiry.	The presentation of experienced facts.	Verifying factual statements by scientific method that stress replicability and universal 'truth'.	The investigation of personally and socially constructed meanings and discourses in ways that stress subjectivity.	The construction of theories that relate observations and experiences to underlying structures and processes.	The construction of critical theories in conditions of free and open dialogue that allow all claims to knowledge to be fairly tested for truth.	The building and testing of hypothetical models of how real processes shape events that we may or may not experience.

Figure 2.1   Six philosophies that yield conceptual frameworks for understanding the relations between society and nature

To begin to answer such questions we need to acknowledge that geographers, like philosophers, do not agree on what things can be said to exist and how knowledge of these things can be produced. We can make sense of their debates and differences by recognizing that a philosophy contains an ontology, an epistemology and a methodology. The **ontology** specifies what it is possible to know: the reality that exists and how it does so. The **epistemology** specifies how that reality can be known: the criteria for judging the truth of a statement about reality. The **methodology** sets out an associated set of rules and procedures to guide research or geographical enquiry (Johnston, 1989a). Figure 2.1 summarizes six of the philosophies that are used by geographers to understand the relations between nature and society. This chapter will make reference to these and suggest that a critical realism, that builds upon dialectical materialism and systems-thinking, provides the most appropriate philosophy for uniting the natural and social sciences. It can help us to understand better the relations between nature and society, and so realize more sustainable forms of development.

2.3 Materialism

There are many ways of explaining a thunderstorm. The two main ways are to regard it as an act of a god, spirit or supernatural being or to regard it as the result of natural forces in the atmosphere. The first type of explanation is idealist, suggesting that there is a higher spiritual reality in terms of which material reality is (at first or last resort) to be explained. The second type is materialist, setting aside preconceived ideas about things and tracing all change back to material causes. The two explanations have different practical implications. The idealist would pray or trust to fate while the materialist might have sufficient knowledge and foresight to build a lightening conductor.

Cornforth (1987) suggests that we can set out the basic tenets of **materialism** (and the opposing tenets of **idealism**) in this way:

- The world is by its very nature material. Everything that exists comes into being on the basis of material causes and develops in accordance with the laws of motion of matter. (The material world is dependent on the spiritual.)
- Matter is objective reality existing outside and independent of the mind. Far from the mental existing in separation from the material, everything mental or spiritual is a product of material processes. (Spirit, mind, or idea, can and does exist in separation from matter.)
- The world and its laws are knowable. While much in the material world may not yet be known there is no unknowable sphere of reality that lies outside the material world. (There exists a realm of the mysterious and unknowable, 'above', or 'beyond', or 'behind' what can be ascertained and known by perception, experience and science.)

Notice that materialists do not deny the role of ideas in shaping our perceptions and actions in the world, nor the possibility of science revealing design or purpose in the material world which idealists may associate with religion or spirituality. The world we perceive and experience is always interpreted or mediated by culture (language, ideas, values), but it does have real, material existence that is independent of people's knowledge. Such **realism** suggests that some (materialist) ideas are more realistic, truthful and useful than other (idealist) ideas.

Materialists believe that people's ideas reflect their material existence in the world and once formed, act back to shape their behaviour and environment. This leads them to associate the origins of idealism with the helplessness and ignorance of people in the face of the forces of nature. Idealism takes many forms, including religion. All involve a doubling of the world (God and man, soul and body, heaven and earth, nature and society, values and facts) and can prompt different approaches to the social and natural sciences if only the former are perceived to be influenced by human subjectivity.

2.4 Premodern thinking about society and nature

Ancient cultures considered the cosmos to be a living organism of which they were a part. The Earth was alive and intelligent like a vast animal or mother, with a soul and rational mind of its own. Tribal peoples and hunter-gatherers understood human society by reference to plants, animals and other natural entities (**totemism**) and modelled nature on prevailing social relations (**animism**). Chapter 3 explains how such beliefs reflect and reinforce the dialectical relationship whereby society and nature mutually constitute one another and it is not possible to understand one without reference to the other. They gave rise to a range of rituals, ceremonies and sacred places that enabled people to meet and communicate with the spirits or gods which were thought to resided in and/or control nature, and also acted as cultural constraints on the exploitation of natural resources. Some contemporary environmentalists who are members of alternative or 'New Age' movements, seek to revive such idealist beliefs and continue to celebrate ancient ceremonies at such places as Glastonbury and Stonehenge.

At the time of the Renaissance in Western Europe people believed the cosmos to be a living organism composed of four elements (earth, air, fire and water) below the moon, and a fifth element (ether) that made the stars and planets (Merchant, 1992). The soul of the Universe caused the movement of Sun, stars and planets around the Earth, and God's spirit descended from the heavens to mingle with the ether and the air and sustain living things. The Earth was considered a receptive nurturing female with circulation (water cycle), bowels (volcanic features), skin (soil) and tresses (trees), and those seeking to exploit or mine the Earth were expected to offer sacrifices or propitiation. Medieval science supported this cosmology, combining the ideas of the Greek philosopher Aristotle with elements of Christian theology. Aristotle saw matter as containing potentiality that is realized by means of form. Matter and form are linked through the process of development in an organic and living universe, and by studying this process science can reveal God's underlying purpose and design. Medieval science embraced God, the human soul and ethics, and by ordering God, people and the rest of nature in a hierarchical manner, it continued to subject resource-exploitation to moral constraint. Such ideas were challenged by the Reformation and by those thinkers who initiated the Enlightenment.

2.5 Modern thinking about society and nature

Sixteenth-century reformers, such as Martin Luther and John Calvin, believed that spirituality or divine purpose was latent only in people and not the rest of nature, and so discouraged belief in such things as holy water and sacred places. Nature was then further desacralized by the rise of science in the sixteenth and seventeenth centuries, with new discoveries in physics, astronomy and mathematics reducing the workings of the world to the mere mechanical interaction of bodies or particles. The theories of Copernicus, Galileo, Descartes, Bacon and Newton (see Box 2.1), were among those that established **mechanical materialism**: the philosophy that accompanied the Age of Enlightenment and facilitated the rise of industrialism, capitalism, liberalism and modern society (see Chapter 5). It generated knowledge of how things worked by basically taking them apart and reassembling them (Cartesian method). This knowledge could then be used to change, perfect and repair nature in ways that promised human progress and development. The Enlightenment of the late seventeenth and early eighteenth century replaced belief in the authority of God with modern belief in the authority of science and human reason (Gaarder, 1995; Osbourne and Edney, 1992).

Mechanical materialism perceives the world as a dead machine, a system in which all matter behaves according to strict mathematical principles and everything is predictable. It requires the scientist to stress substance over form; value the measurable over the immeasurable; separate facts from values; place people outside nature and tradition; and specialize on one part of reality. Furthermore it provides no adequate material explanation of origins (and hence opens the door to idealism); reduces all phenomena to the workings of a machine with no capacity to change or evolve; and cannot reveal the real material economic, political and cultural processes that explain social development.

Despite these weaknesses, mechanical materialism prompted a similar approach to the natural and social sciences. Social science developed in ways that mirrored natural science in its attention to individuals, their functions within society seen as a smoothly working machine, and the ways in which their behaviour might be modified and improved. **Positivism** (see Figure 2.1) is the dominant philosophy shaping modern science and modern approaches to understanding and managing the environment. It shapes technocentric environmentalism (p. 19) and several of Dryzek's environmental discourses, including sustainable development and ecological modernization (Figure 1.8).

Box 2.1   Some early modern scientists and philosophers

Nicholaus Copernicus, 1473–1543
Shattered the static medieval world-view by picturing the Earth circling the Sun. He tried to verify his theory by observation.

Galileo [Galilei], 1564–1642
Confirmed Copernicus' theory using the new telescope and worked out the fundamentals of a theory of motion.

Francis Bacon, 1561–1626
Developed science as an experimental method for manipulating and gaining knowledge of nature. Thought that nature should be 'bound into service', made a 'slave', put 'in constraint' and 'moulded' until she gave up her secrets. The Royal Society was founded in 1660 to further Bacon's aims of using science to advance human progress.

René Descartes, 1596–1650
The father of modern rationalism suggested that the foundation of all knowledge is reason. He outlined four rules for clear reasoning: (1) never accept anything except clear and distinct ideas; (2) divide each problem into as many parts as are needed to solve it; (3) order your thoughts from the simple to the complex; (4) always check thoroughly for oversights. He believed that the essence of being is thinking, and that the mind is separate from the body (Cartesian dualism).

John Locke, 1632–1704
Opposed Descartes' rationalism by asserting that all knowledge is based on experience (empiricism). Locke considered the mind of the newborn child to be like a blank sheet of paper. It acquires simple ideas through sensory experience and by combining, comparing, or abstracting from simple ideas to produce more complex ideas. All ideas therefore have their origins in the qualities of objects and the primary source of knowledge is sensory experience.

Isaac Newton, 1642–1727
Using his new mathematical tool, the calculus, Newton linked Galileo's mechanics with Keplar's planetary laws and Gilbert's theory of attraction to produce a grand universal scheme: a mechanical Universe that still dominates the popular imagination.

Immanuel Kant, 1724–1804
Overcame rationalism's problem of linking rational certainty with reality and empiricism's problem of providing the logical necessity of laws based on sensory experience. He suggested that rationalism deals in analytic propositions (categories of thinking that enable us to speculate on such phenomena as the laws of gravitation), while empiricism produces synthetic propositions (knowledge that allows us to verify such laws at work).

Modern knowledge displays three basic characteristics (Pilkington, 1997):

- **Foundationalism** There are indisputable foundations for knowledge in sensory experience (empiricism) and rationality (rationalism). Scientific enquiry and reason can reveal the essential truth about the world.
- **Totalization** It is possible to advance general or universal theories about nature, society and history. They each have an inner logic and are ordered according to universal laws.
- **Utopianism** The application of increasing knowledge brings constant improvement in the human condition. Science, technology and bureaucracy offer rational control of nature and society, and thereby will bring material prosperity, individual liberty, social equality, universal morality and emancipation from natural calamity, poverty, disease and political oppression. This is sometimes referred to as the modern project.

2.6 Romanticism and organicism

Mechanical materialism was opposed in the late eighteenth and nineteenth centuries by a romantic movement in art, literature and philosophy. **Romanticism** rejects the kind of rationalism and progress associated with modern science, arguing that colour, taste and feeling are as important as shape, number and movement, and that distinctions between reason and emotion, or between mind and body, are misleading. Romantics such as Blake, Wordsworth and Thoreau stressed the realms of the subjective (imagination, feelings); preferred rural simplicity and innocence to urban complexity and sophistication; and expressed a spiritual affinity for nature as purifying people's souls and expressing the divine (Figure 2.2). Following Kant and Schelling, they believed that reality is ultimately spiritual (idealism). Nature is the mirror of the human soul, a creative spirit whose aspiration is ever fuller and more complete self-realization.

Romanticism is holist: perceiving the world as an harmonious and self-organizing whole that can only be understood in its totality by stressing wholes over parts, form over substance, and change over mechanical equilibrium. It correctly insists that morality cannot be excluded from science, but idealizes an imaginary nature and 'natural condition' separate from people and society. Such romantic idealism continues to influence those ecocentric environmentalists, or green romantics, who suggest that we can somehow 'return to nature' or live 'according to nature'.

Meanwhile mechanical materialism and Cartesian method influenced the biological sciences as medicine treated the body as a machine and biology, aided by the invention of the microscope, took analysis down to the level of cells and genes in its search for fundamentals. Cell theory, embryology, microbiology, heredity and evolutionary thought all reinforced mechanistic conceptions of life, but these were challenged by **organicism** that suggested that living things were not merely functioning machines but patterns of relations capable of self-organization and development. In the early twentieth century organismic biologists pointed to the outstanding tendency of all life to form

Figure 2.2   Romanticism is reflected in the eighteenth-century English landscape garden. Designers such as William Kent and Capability Brown set elements of the picturesque within a larger idealized natural landscape
Source: TRIP/J. Merryweather.

multilevelled structures of systems within systems (cells, tissues, organs, organisms, ecosystems, social systems) with different levels of complexity and different laws operating at each level (Capra, 1997). Their systemic thinking was echoed by that of social scientists who were developing dialectical materialism.

2.7 Dialectical materialism

Dialectical materialism addresses the weaknesses of mechanical materialism by understanding the world not as a complex of ready-made things or particles, but as a system of processes, flows and relations through which all things come into being, exist and pass away. Particles, things, structures and systems do not exist outside of or prior to the processes, flows and relations that create, sustain, or undermine them (an ontological principle). All are constituted by matter in motion: matter driven by physical, biological and social processes that are made possible by the relations between the things that make up the biophysical and social worlds. Flows of energy, material and information, within and between these worlds, create, sustain and undermine human environments, and should be the primary focus of the environmental geographer's search for knowledge and truth (an epistemological principle).

Dialectical materialism overcomes the weaknesses of mechanical materialism by maintaining that matter and motion have always existed; that change is part of the very nature of matter with inner impulses of development inherent in all things; and that social development results from the contradictory nature of the processes, flows and relations shaping society and its environment. Mechanical materialism's single process of mechanical interaction between objects is replaced by a great variety of evolving processes as simpler forms of matter in motion are continuously transformed into higher states. Life evolves at a certain stage in the development of more complex forms of non-living matter. Life takes human form and some human societies develop to a stage where they exploit the biophysical world in non-sustainable ways. Contradictions between the promise and reality of modern development may then challenge people to develop new processes, flows and relations in society (new technologies and forms of social organization) which put development on a more sustainable path.

Mechanical materialism assumes that things have their own fixed nature; that they exist in isolation independently of everything else; and that the whole is merely the sum of its parts. Dialectical materialism rejects these assumptions claiming that nothing exists in isolation; that things can only be understood in relation to other things; that parts and wholes are mutually constitutive of each other; and that the whole is more than the sum of its parts. The individual and society, part and whole, both change as people form new and lasting relations in the home, workplace, state and elsewhere. People are both the subjects and objects of social change in the same way as organisms are the subjects and objects of evolution (Harvey, 1996).

Dialectics suggests that the environmental systems that people create are always contradictory and problematic because of the multiple processes which shape them. Opposing forces are always at work and produce creative tensions from which transformative behaviour can arise. You might like to consider what contradictions and tensions were revealed by the two case-study boxes in Chapter 1 (the history of the Marston Vale and the contemporary state of the world) or what contradictions and tensions are currently present in your own environment. This should help you recognize that change and instability are the norm and that stability or equilibrium in systems may be harder to explain. The **four principles and three levels of dialectics** are outlined in Figure 2.3.

DIALECTICS seeks to explain the general laws of movement at the levels of nature, society and thought and reflects four principles:

- TOTALITY or everything is related. Nature is a coherent whole. Things are related and reciprocally condition each other. They are to be understood in their concrete totality. Everything has something to do with everything else. Nothing is isolated.
- MOVEMENT or everything is constantly being transformed. Nature is in a state of becoming. Movement is a quality in everything. Nature, society and thought are not fixed but continually being transformed, never definitely established, always unfinished. The cause of movement is internal struggle or contradiction. The general movement of reality makes common sense. It accords with our experience of the world as one of becoming, existing and passing away.
- QUALITATIVE CHANGE or the tendency to self-organization and complexity. Transformation in nature is not a circular process of endless repetition but an evolutionary process towards higher states of self-organization and complexity.
- CONTRADICTION or the unity and struggle of opposites. The transformation or evolution of things is only possible because opposing forces coexist within them and simultaneously move toward unity and opposition. Such contradiction is inherent in all things (nature, society and thought) and is the cause of movement whereby contradictory aspects may attain a higher state of resolution (organization, complexity) that is always conditional, temporal, transitory and relative.

Figure 2.3 The three levels and four principles of dialectics
Source: Gadotti, 1996, pp. 17–20.

2.8 Historical materialism, political ecology and critical theory

It was Marx and Engels who used dialectical materialism to understand people's development within the rest of the natural world. Marx sought to understand history not in terms of substance (as mechanical materialists had done), nor in terms of changing consciousness (as idealists like Hegel had done), but in terms of the changing relations between people as they work to produce their needs and wants (Slaughter, 1985). **Historical materialism** regards work as the key process of mediation between society and nature: a process whereby people transform nature and are themselves transformed. As we saw in Chapter 1, work in the broadest sense shapes our metabolism with the rest of nature; converts first nature (untouched by people) into second nature (the environment and culture); and can enable individuals and societies to realize higher states of personal and social development. Whether or not this happens depends on the relations between people (social relations) and between people and the rest of nature (environmental relations). Such relations are about ownership and control, and if they are unequal and undemocratic – as they are under slavery, feudalism and capitalism – then human and non-human nature are exploited in the interests of a minority, and unsustainable forms of development are the result (Croall and Rankin, 2000; Peet, 1991).

Marxists interpret history as a sequence of **social formations** each using **forces of production** (land, labour and capital) in different ways, with different social relations, institutions and systems of beliefs and values. In modern capitalist societies the key contradiction is between private property and the social character of production. Individuals are separated or alienated from external nature (the rest of the biophysical world) and their own inner nature (their potential as human beings) by work which separates them from the product of their labour; by living conditions that separate town and country; by consumption patterns that fail to reveal the true environmental costs of goods and services; and by education and media that fail to provide a coherent understanding of the relations between the biophysical and social worlds. Marx believed that democratization of social and environmental relations under socialism would end such **alienation** and enable people to realize their common interest in more sustainable forms of development.

Marx's ideas have been developed as **political ecology** that combines ideas from Marxist political economy with ecology in recognizing both the first and second contradictions of capitalism (O'Connor, 1988; Walker, 1998). The contradiction between the forces and relations of production prevents economic development that meets everyone's needs, and is the focus of workers' movements and the old class politics. That between the conditions of production (natural resources and services, human health and education, and rural and urban space) and the social relations that reproduce these conditions, prevents environmental protection and the safeguarding of people's quality of life, and is the focus of the new politics and social movements. As economic crises, caused by capitalism periodically producing more than markets can consume, are compounded by environmental crises, environmental politics reflects old and new interests that support different programmes for more sustainable forms of development. You may like to compare the political ecologists' view of crisis with that revealed by the four-capital model of environmental economics cited in Chapter 1.

Key questions for Marxists are why the contradictions of capitalism have not prompted radical social change and why actually existing socialist societies have failed to deliver social justice and sustainability? Critical theorists, such as Herbert Marcuse and Jurgen Habermas, suggest that in this century, capitalism has developed new compensations and forms of social control (Kellner, 1989). While consumer society, the welfare state, representative democracy, and the mass media, serve to reduce alienation and disaffection, more and more aspects of social life, including the economy and government, operate according to **instrumental rationality** based on positivism. The imperatives of what is technically and bureaucratically expedient now take precedence over what is morally and politically right, with technocrats and bureaucrats taking more and more control over such policy areas as energy, transport, health and education. Technocracy reached extreme forms in the societies of Eastern Europe and the USSR, and the resulting economic and environmental crises eventually prompted their reform.

Habermas believes that this dominance of positivism, instrumental reason and technocracy, currently arrests human progress. It will only be resumed when movements within civil society, including sections of the environmental movement, succeed in establishing more radical forms of democracy governed by **communicative rationality**. This would allow sustainable development to be based on considerations of what is technically possible, culturally appropriate and morally and politically right, that would take into account the interests of present and future generations and those of the rest of the living world (Dryzek, 1996).

2.9 Systems-thinking and complexity theory

While Marxist social scientists were developing dialectical materialism, some natural scientists continued to develop systems-thinking or the understanding of phenomena within the context of a larger whole. Developments in quantum physics, ecology and psychology reinforced those in organismic biology by suggesting that the essential properties of systems are properties of the whole and that systems cannot be understood by the kinds of analysis associated with mechanical materialism. Ecology emerged out of organismic biology in the nineteenth century as biologists began to study communities of organisms. In 1866 Ernst Haeckel defined ecology as 'the science of the relations between the organism and the surrounding outer world' but the term '*umwelt*' or environment was not used until 1909 when it was introduced by Jakob von Üxkull. Ideas about food-chains and ecosystems, introduced by Elton and Tansley in the 1920s and 1930s, further shifted the focus from organisms to ecological systems, reviving premodern ideas of a network or web of life, and helping geographers to develop new understandings of the biophysical landscape.

Systems-thinking reinforces and develops dialectical materialism, suggesting that reality is material, relational, contradictory and systemic (Wu, 1996). All matter exists and develops in the form of systems, and such development is governed by universal relations such as those that determine the chemical composition of the universe, the unity of mathematics, or the theory of relativity. Systems and relations result from the continuous exchange of matter, energy and information, and it is the processes of exchange, between the elements in the system and the system and its environment, that reveal its structure and function. While the second law of thermodynamics suggests that the universe tends towards disorder, dissolution and decay, a contradictory tendency continues to produce systems displaying order, structure and organization.

General systems theory and **complexity theory** has been developed with the aid of computers to understand the systemic nature of the world (Lewin, 1997). Most of the systems of interest to the geographer (weather, ecosystems, economic systems, cultural systems) are non-linear and do not work in the 'clockwork manner' assumed by mechanical materialism. Although they display huge numbers of elements, relations and exchanges, the richness of the interactions allows the system as a whole to undergo spontaneous self-organization. Genes in a developing embryo organize themselves to form a liver cell; organisms constantly adapting to each other through evolution organize themselves into an ecosystem; people trying to satisfy their material needs unconsciously organize themselves into an economy. Elements seeking mutual accommodation and self-consistency somehow manage to transcend themselves, acquiring collective properties such as life, thought and purpose, that they might never have possessed individually. Mathematical tools and ideas from such fields as artificial intelligence, ecology and chaos theory, have allowed complexity theory to grasp the self-organizing dynamics of the world in new ways, heralding an alternative kind of science for postmodern times.

Self-organizing systems are adaptive, producing a compression of information with which they can predict the environment. DNA encodes information for biological evolution, while cultural institutions, customs, traditions and myths encode information for social evolution. Complex adaptive systems evolve to the 'edge of chaos' where order and chaos are finally balanced, the capacity for information storage and processing is maximized, and there are strong prospects for continued creativity and co-evolution with other systems. Such states prompt the kind of breakthrough to higher forms of organization that is currently required if social systems are to learn to live sustainably with biophysical systems.

2.10 Postmodern ways of thinking about society and nature ____

Many commentators claim that changes in material and cultural life since the 1960s have been so significant that we are now living in societies that have moved beyond modernity. They argue that postmodern society (postmodernity) displays such characteristics as segmented production and consumption; mass communications and information technology; globalization and confusions over time and space; new social polarizations and forms of identity politics; the rise of new cults and fundamentalisms, and a greater tolerance of minorities and difference (Jencks, 1996). Chapter 9 will explain these characteristics as products of capitalism's recent development, but for the moment it is important to note that theories of postmodernity emerged in response to a loss of faith in the modern project. Postmodernity is associated with artistic and intellectual styles (postmodernism) which build on **interpretivism** (see Figure 2.1) and seek to deconstruct the characteristics of modern thought outlined above. Postmodernism says 'no' to all the foundational principles of modernism: principles that seek to make the world knowable, accountable, unambiguous, generalizable, predicable, manageable and mutually comprehensible. The ideas of some postmodern philosophers are outlined in Box 2.2 and you will notice that as in Box 2.1, these are all men! You might like to research introductory texts on philosophy in an attempt to restore the balance. Among around a hundred men, Osbourne and Edney (1992) mention only two women: Mary Wollstonecraft and Rosa Luxemburg.

Postmodern knowledge displays three basic characteristics (Pilkington, 1997):

- **Anti-foundationalism** There are no indisputable foundations for knowledge; no general criteria to distinguish truth and falsity. Language, thought and reality are interdependent and all knowledge is mediated through language rather than being an accurate reflection of nature. Truth is relative and there are no guarantees of truth or reality outside language or discourse.
- **Anti-totalization** It is arrogant to advance general theories that pretend to reveal universal truths or meanings. We should abandon such attempts and accept a diversity of limited theories and truths. We should be particularly

Box 2.2   Some twentieth-century postmodern philosophers

Jean-François Lyotard

Something funny is happening to grand narratives (the big stories of science, progress, Marxism, humanism, etc.). They are fragmenting into little, local stories and struggles with their own irreconcilable truths.

Michel Foucault

Power has always worked in dispersed ways, through the everyday practices of our institutions, and in the ways in which we inspect, name and blame one another. Power lurks and works in discourse. It is tied to knowledge and language and so are 'we'. The human subject is the contested and shifting creature of different discourses.

Jean Baudrillard

Language has gone 'hyper'. Words and images no longer refer to anything except one another as they hurtle around electronic information and entertainment networks. We no longer have access to what is real. We have become passive spectators of events without consequence.

Jacques Derrida

Language is all. There is no point in looking outside language for the truth that lies behind it, or the thought that it expresses, or the reality that it reflects, or the objects which it represents. There are no first, last or deepest things out there. There are no foundations to ground our theories, justify our politics, validate our beliefs, centre our selves. We should attend to what happens between and among signs because that is where meaning lies. Much philosophy, science and logic sets up pairs of terms which privilege one of the pair and exclude the other. The task is to deconstruct those oppositions, and challenge the closure of meaning and possibility that they bring about.

Source: Maclure, 1995

sceptical of totalizing thinking that seeks to explain the world from centred and privileged positions of male power.

- **Anti-utopianism** Modern knowledge has not delivered utopia or enlightenment, but has resulted in oppression and domination. There is no justification for accepting grand stories or narratives of human progress that suggest that history has purpose and that things will continually get better.

These characteristics are reflected in a consumer and media culture which offers people multiple meanings and identities, and an academic culture which is preoccupied with the deconstruction and reconstruction of texts and discourses of all kinds. Postmodern geography focuses on the social construction of nature, the environment and environmental issues and risks, in such texts as conversations, newspaper reports, textbooks, television documentaries,

Figure 2.4 What meanings of nature are constructed for and consumed by these tourists in Kenya? A research question for the postmodern geographer
Source: TRIP/D. Saunders.

advertisements and theme parks (Figure 2.4). It seeks to deconstruct 'expert' accounts and reconstruct the environment and the rest of nature as significant partners whose numerous voices, stories and meanings can inform our decision-making. This requires the attention to subjective experience and the role of discourse in everyday sites of biological, social and cultural reproduction that is reflected in much feminist writing.

Postmodern approaches to environmental planning reflect a shift from mechanical materialism, positivism, modern science and technocracy, to dialectics, holism, citizens' science, and radical democracy (Dyck, 1998). Postmodern environmental designs draw on a wide range of academic and local knowledge, involve new forms of community participation, and seek co-evolution of social and biophysical systems in ways which respect cultural and ecological integrity. They are more modest in their ambition and scope and reflect democratic pragmatism (Figure 1.8, p. 20) by involving networks of stakeholders in developing and practising new forms of environmental politics. Such designs can usefully draw on an understanding of critical realism.

2.11 Critical realism

The need for more sustainable forms of development challenges humanity to reflect and act on a philosophy that can hold together nature and society and provide insights into how social systems should evolve alongside biophysical systems. Critical realism provides such a philosophy since it is able to incorporate dialectical materialism, critical theory, systemic thinking and postmodernism, while avoiding the threat of idealism and moral relativism inherent in some postmodern ideas (Cloke, Philo and Sadler, 1991; Dickens, 1992, 1996; Sayer, 1985).

Critical realism recognizes three levels of abstraction at which things can be examined and knowledge generated. At the deepest or more abstract level are the real objective powers of objects or processes made possible by relations. At an intermediate level are more contingent factors, specific to given historical and social circumstances, which determine whether or not objective powers are realized (whether processes cause events). At the surface level are experienced phenomena which arise out of the combination of objective powers with contingent factors, and can be observed at a given place and time. Realist explanation consists of connecting experience in the empirical domain to structures and processes in the real domain through contingent factors in the actual domain.

Realism offers a unified approach to the natural and social sciences while recognizing real but different structures and processes within the physical, biological and social worlds. The biological world is emergent from the physical world, and the social world emergent from the physical and biological worlds. The new physical and life sciences enable us to understand the dialectical and systemic nature of the physical and biological worlds and the processes of emergence whereby the Universe shows ever increasing self-organization and complexity.

Social science needs to be combined with natural science to understand how society is embedded in nature, while natural science needs to be combined with social science to understand the forms which nature takes in specific historical and social circumstances. Realism offers a unified science but acknowledges that there are limits to naturalism (studying society in the same way of nature) which derive from the distinctive emergent properties of societies. Social structures, unlike natural structures, do not exist independently of the activities they govern (they exist only by virtue of the activities they govern); do not exist independently of people's conceptions of what they are doing in their activity; and are usually only 'relatively enduring'. The processes that they enable are not universal or unchanging over time and space.

Chapter 1 explained that the forms of nature which people experience (their bodies, other people, food, landscape, the countryside, urban environments) are socially (culturally) constructed in two senses: they are materially shaped by social practices *and* they are experienced through the mediation of cultural discourses and representations. But we need to be careful of the term 'construct'. Such natures are not wholly artefacts of society or culture since they are in part shaped by real structures and processes in the biophysical (natural) world. It is to these structures and processes that the **realist concept of nature** refers (p. 5).

People have powers that they realize by establishing relationships with the rest of nature, including other members of society. Their interactions with the natural world go beyond mere survival and reproduction for in appropriate circumstances they are able to develop their sociability and their intellectual, aesthetic and spiritual potentialities. People are part of nature (have **natural powers** like other animals), yet are apart from nature (have unique species powers), and can only realize their powers when they are not separated or alienated from nature by oppressive relations. Their potential for applying their growing understanding of the world to the regulation of these relations provides grounds for optimism concerning progress towards more sustainable forms of development.

As we saw in Chapter 1, nature in a realist sense sets elastic limits on how people can live in the world. They themselves must decide what forms of ethics, politics and governance should regulate their relations with the rest of human and non-human nature. Environmental ethics emerge once people realize that the world they inhabit is their own construction and responsibility, and start turning the actions whereby they construct nature into the objects of explicit and discursively justified communal choice. Sustainability is a discursively constructed concept that is most likely to incorporate inter- and intragenerational equity if it is constructed in communities displaying radical forms of democracy and citizenship.

Critical realism can incorporate postmodernism's attention to discourse, subjectivity, pluralism, the new science, co-evolution and radical politics, but represents an approach to knowledge that lies between those of modernity and postmodernity. This position can be summarized as follows:

- **Foundationalism/anti-foundationalism** Acknowledging the inevitable cultural mediation of reality does not mean that there are no criteria at all, or indeed general criteria, for deciding what is true or right. Nor does it mean that there is no connection between language and discourse and the real world.
- **Totalization/anti-totalization** Accepting a multitude of limited theories or texts should not mean abandoning the search for general theories that seek to show how these are related to one another.
- **Utopianism/anti-utopianism** Modern rationality can be used to dominate or liberate. The problem is not the modern notion of progress but its partial realization. We should retain a realistic utopianism.

2.12 Studying society and nature

This chapter should have convinced you that there is no one way of studying nature, the environment or environmental issues. Dialectics and critical realism claim that knowledge and truth are practical questions, or that the validity and power of ideas is demonstrated by their utility. Knowledge starts from activity in the material world and returns to it dialectically. Theory is a guide for practice and practice a test of theory. People are beings of praxis (reflection and action) and it is through social learning or revolutionary praxis that they can overcome their alienation from nature, realize their species powers, and make the transition to more sustainable forms of development. Education should not be a product or body of irrelevant knowledge to be taught and learned, but a form of praxis whereby teachers and students create their own socially useful knowledge by reflecting and acting on the events and issues they experience from day to day (Gadotti, 1996). Participation in community efforts to realize sustainability should form a significant element of courses in environmental geography, for it is only in such settings that the true value of academic ideas can be evaluated. Making sense of your environment, and yourself as a player in that environment, requires both theory and practice if you are to develop a realistic vision of a more sustainable future and the means whereby it can be realized.

2.13 Further reading

Engels wrote his *Introduction to Dialectics of Nature* in 1875–76. Read this essay (Engels, 1950) and then read Chapter 2 of Harvey (1996). In what ways does dialectics challenge your ways of thinking about nature and the environment and the geography you have been taught to date?

There are few simple introductions to critical realism but you may find much of Dickens (1992; 1996) and Collier (1994) accessible. Proctor (1998) examines critical realist responses to postmodern relativism. To what extent do you agree with Dickens' thesis that academic divisions of labour contribute to our alienation from nature?

Postmodern ecology challenges the convictions of many environmentalists who believe that nature is constant, eternal and self-regulating except when people intrude. Read Budiansky (1996) and Zimmerer (1994) to find out about the deconstruction of modern ecology, the construction of a postmodern ecology, and the implications of such science for geography and environmental management.

Chapter 3

Premodern environments

3.1 Introduction

Archaeological discovery and Darwinist evolutionary theory will probably have convinced most readers that early hominids evolved from apes between five and ten million years ago and that the human genus (*homo*) appeared in Africa about two million years ago. At about this time, humans first employed technology, in the form of rudimentary stone cutting tools (Goudie, 1990). The emergence of 'modern' society, as characterized by the diminishing authority of religious hierarchies and a gradual acceptance of scientific authority, is associated with a comparably modest duration, occupying the last 500 years. Indeed, if the history of human occupation of the Earth were analogized to a hundred-metre sprint, the race through modernity would only involve the final couple of centimetres. This chapter examines premodern relationships with nature, considering the connection between different productive regimes and their accompanying cultural developments. Understanding the respective roles of nature, culture and technology in both gradual and revolutionary changes to human environments is a fascinating area of enquiry, one which has gripped the imagination of generations of geographers, environmentalists and anthropologists.

3.2 Why study premodern societies?

The study of premodern peoples and their changing environments may be of intrinsic interest, appealing to our curiosity about 'how things were in the past'. However, we also study environmental history because it can inform contemporary debates about environmental change. Studying premodern environments contributes to the following understanding of contemporary environments.

- The study of premodern societies brings the relationship between culture and natural resources into focus. The co-evolution of culture and environment, mediated through human agency, is illustrated through the consideration of technological intervention, economic organization and systems of belief.
- Studying the dynamics of a changing relationship with nature provides insight into underlying *processes* of change – processes which might still be relevant, albeit under different contextual circumstances.

41

- Studying premodern societies may even provide role models for post-industrial society, demonstrating an alternative road to affluence (Sahlins, 1972) or a blueprint for sustainable relationships with nature (Pereira and Seabrook, 1990). You may, for example, have noticed the influence which premodern societies have had on some 'New Age' thinking.
- While modern 'scientific' knowledge has widely come to be considered as a superior replacement for traditional 'indigenous' knowledge, this assumption is currently being re-evaluated. Sensible commentators now look for ways in which different knowledge systems can complement each other, rather than attempting polarized judgements in favour of one or the other. For example, many traditional farming and fishing systems are now being recognized for their intricate understanding of the capabilities of local ecosystems – an understanding which the modern world could well learn from (see Richards, 1985).

3.3 How do we learn about premodern societies?

The earliest written records of human societies date back about 5,000 years, with texts found in the Nile Valley, Mesopotamia and the Indus Valley. Many premodern societies were oral cultures and we have inherited no writings about their ways of life. As a result, geographers receive much of their information from archaeologists and anthropologists. Archaeological finds provide information about technology, diet, spatial dispersion, migration patterns and so on. However, for detailed information about culture we mostly rely on ethnographic studies of remnant hunter-gatherer and tribal farming populations. This generates its own hazards: for the most part remaining hunter-gatherer societies have had significant contact with modern societies and have undergone change as a result. Care is therefore needed when assuming that such groups are truly representative of antiquity.

The cultural variety found in premodern societies is so great as to be literally indescribable. Even to catalogue the different belief systems among tribal people in a small nation like Nepal is considered an impossible task, owing to the constantly evolving pantheon of gods which themselves serve different roles for different village communities at different times of year (Krausskopff, 1996). Attempting to extend such a catalogue across three million years of time and the entire inhabited space of the Earth would be wholesale folly. Nevertheless, while it is important to recognize the remarkable cultural diversity which time and space have conspired to produce, it is also useful to identify certain common features of premodern societies. A search for commonalities allows some generalizations to be made, and these will help to organize our thinking. Generalizations and categorizations must be employed with caution, recognised as the rather blunt conceptual tools that they invariably are. Nevertheless, they must be employed because life would become highly confusing without them. With this brief justification, we can proceed to classify premodern societies into a small number of 'types'.

3.4 Types of premodern society: modes of production, cosmologies and ideologies

For more than 99.5 per cent of human history, the world's population has lived from hunting and gathering alone, and has mainly been restricted to small, nomadic bands or tribes, generally numbering less than 40 members. The cultivation of plants and the domestication of animals, basic features of sedentary agriculture and pastoralism, began around the end of the last ice age, 10,000 years ago. Sedentary agriculture (along with a new-found taste for conquest and plunder) provided the economic conditions under which state-organized civilizations could first emerge some 6,000 years ago in the Middle East. In these traditional states, agriculture remained the prevalent productive activity, but cities also appeared, along with the urban and administrative trappings of what is often referred to as 'civilization'. In this briefest of premodern histories, three types of society become discernible: hunter-gatherer, agricultural and traditional state societies. A mosaic of different physical environments, combined with locally induced innovations and culturally specific decisions, means that these three basic types manifest themselves in a rich variety of local forms.

The commonalities that distinguish these three types of society are not restricted to the technical methods of producing and consuming food. What has so far been called a 'type of society' can more usefully be called a **mode of production**. This is a classificatory tool that is used to describe the common economic and social practices shared by an array of different societies. On the one hand, it describes the fundamental characteristics of the way in which the use of nature is organized by a society (the system of production). On the other hand, it also describes the system of ideas and beliefs that accompanies a particular system of production (see Figures 3.1 and 3.2). In describing the organization of the productive use of nature, it is common to consider the relationship between the **forces of production** (the energy, machines and so on which are used to physically transform natural resources into products for human consumption) and the **relations of production** (the stratification of society by class and gender, determining who owns the forces of production, who does what work and how surplus is distributed). The social relations and forces of production are closely related and can be described as reciprocally constituting each other. As a consequence of this interrelationship, we can state that *changes to social relations will necessarily involve changes to society's relationship with nature,* and vice versa. This statement has enormous consequences for the way in which we should set about understanding and managing environmental change. It establishes a theme which underpins much of this book.

An economic mode of production goes some way towards describing a society and its relationship with a surrounding environment. This description must be made more complete through an examination of a society's social structures. Societies are usually considered to have some mechanisms that can potentially lead to the transformation of the forces and relations of

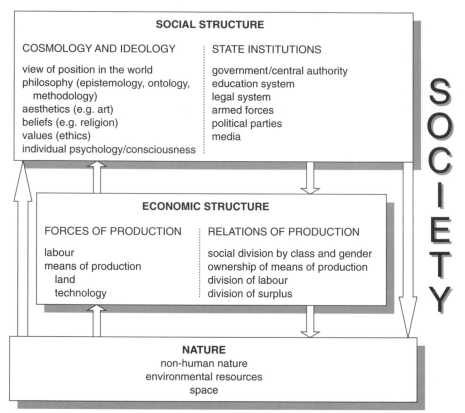

Figure 3.1   The mode of production model of society

The arrows suggest interrelationships between all components. The nature of these relation-ships is of great interest to a wide range of academic disciplines and continues to generate intense debate. The idealist philosophy of Hegel considered consciousness (in the realm of cosmology) to be the prime mover, being the instigator of change. This contrasts with the materialistic philosophy of Marxism which considers the economic realm to be the prime mover. The early work of Karl Marx tended to assign primacy to the forces of production, while his later works considered the relations of production to be the dominant partner (Hindess and Hirst, 1975; Peet, 1998). Atkinson (1991) asserts that environmentalists are either 'baby Hegels' or 'baby Marxes', although sophisticated philosophical positions attempt to transcend such polarized thinking. For some geographers of the early twentieth century (the so-called 'environmental determinists'), nature was the prime agent, determining the possible forms of production and associated belief systems. Ian Simmons, a geographer at Durham University, has strongly advocated the importance of technology in driving socioenvironmental change.

production. In particular, it is possible to transform the kind of technologies employed and the way in which resources and profits are distributed. As noted in Chapter 2, sociobiophysical systems tend to be far from equilibrium: pro-cesses of change are always operating and systems tend to evolve into more complex, emergent forms of organization (Figure 3.2). In order for an existing mode of production to resist forces of change (to achieve a semblance of equilibrium), there must be a dominant ideology that leads to the belief that

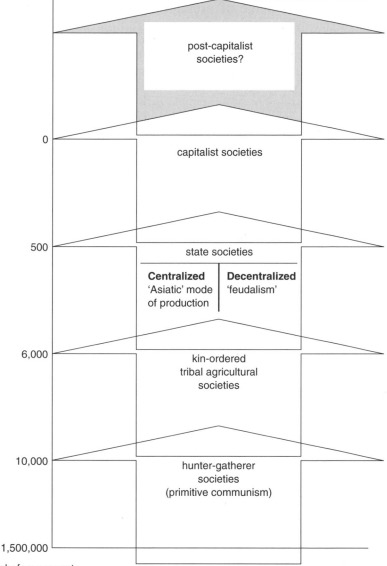

Figure 3.2 Modes of production through time
Source: Peet, 1998, p. 84 (with 'post-capitalist' arrow added).

those forces and relations of production are the best available. The existing system of production must also be suitably compatible with a system of held beliefs and values. In later societies, the state acts as the guardian of existing social and economic structures, reinforcing them through political and legal systems, education and force. In hunter-gatherer societies, the forces against change are held in a system of communal beliefs related to a **cosmology**: a system of knowledge about how the Universe operates.

3.5 Hunter–gatherer mode of production

Humans have lived through a combination of gathering and hunting for the great majority of their lived history. Archaeological evidence suggests that humans first evolved in East Africa and had occupied the entire inhabitable Earth some 35,000 years ago (Simmons, 1989). This geographical spread demonstrates the capability of a hunting and gathering mode of production to operate successfully in a wide range of habitats. Today, only remnant bands of these societies are left, mainly in remote mountains, forests, deserts and polar regions. It has proved virtually impossible for this mode of production to survive the confrontation with a modern society that has proved to be a consistently threatening and intolerant neighbour.

The forces of production are fairly simple in hunting and gathering societies, involving only rudimentary stone and wooden tools for carrying, hunting, digging and cooking, and relying on the Sun as the only source of energy (in the form of energy harnessed in plant and animal growth). However, the use of nature often became quite complex. Groups commonly move about within their environment, exploiting seasonally available resources in different places and spreading the harvesting of resources across a sufficient area to avoid exhaustion of the resource. Hunting and gathering regimes required an accumulated knowledge of seasonality, breeding patterns, migration, plant life-cycles and so on. In North America, bands of hunters were capable of prolific feats, slaughtering up to 200 buffalo at a time by driving them over a cliff or into a narrow canyon.

Simmons (1989, p. 40) describes the use of fire as 'the first great revolutionary agent in mankind–environment relations'. Perhaps the first impact of the control of fire was the ability to colonize cooler climates. Later, North American and Australian hunter-gatherers used fire as an aid to hunting and, less commonly, as a means of altering vegetation structures in favour of more desirable species or products. In Britain, the first evidence of domestic fire use is found 400,000–500,000 years ago at Westbury Sub-Mendip (Simmons, 1989) while the last hunter-gatherers in England and Wales (9,500–5,500 years ago) used fire to clear settlements, which sometimes led to permanent conversion of upland forest into moorland (Simmons, 1997).

Hunter-gatherer societies are consistently characterized as having egalitarian social relations. Broadly speaking, relations of production that involve a fair distribution of surplus can be described as egalitarian, while relations of production which involve the unequal appropriation of surplus by a dominant class or gender might be described as inegalitarian. In hunter-gatherer societies, there is little difference in status among band members. Surplus time (i.e. that not needed for personal and cultural survival) is not spent in creating wealth for a particular class of people, and individuals do not therefore amass personal possessions. For one thing, the hoarding of goods makes no sense. The need for mobility, combined with the inability to harness animal power for transport, ensures that there is no desire to own more than can be comfortably carried. Cooperation is the norm in these societies, and hunting,

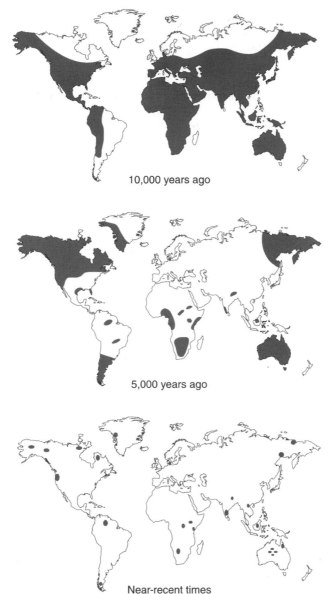

10,000 years ago

5,000 years ago

Near-recent times

Figure 3.3   Distribution and decline of hunter-gatherer societies
Source: After Lee and de Vore (eds), 1968.

gathering and religious activities are performed for the group rather than for individuals. While some behavioural psychologists will hold that the lengthy hunting heritage can be viewed as an explanation for the persistence of violent tendencies within contemporary society, this seems to be misleading. Hunters must often carry weapons that are capable of taking human life. However, instances of violence (and especially interband warfare) seem rare and loss of life appears to have been minimized. The relations of production in hunting societies do tend to include a division of labour by gender. In particular, the use of weapons to hunt larger game usually renders hunting an exclusively male occupation: a mechanism to protect the reproductive role of women. In addition, elder men generally have a special role to play in major decision-making. However, there remains relatively little power differential and all adults participate in decision-making (Giddens, 1989). For this reason, some geographers, such as Richard Peet (1991; 1998), adopt the Marxian term 'primitive communism' to describe the hunter-gatherer mode of production.

The revolutionary changes that periodically transform modes of production (Figure 3.2) are nearly always considered to be progressive, especially when viewed from our current position of relative material affluence. But while we should be careful not to idealize hunting and gathering (and certainly not to suggest a mass return to this way of living!), it is worth thinking about what has been sacrificed in the name of progress: a society in which there is little physical violence; no rich and poor; no weak and powerful; no rulers and ruled – a society in which decision-making is cooperative and communal, where all production serves the common good, where productive activities tread relatively lightly on the surrounding environment and where food is sufficiently abundant to enable the majority of time to be spent at rest (see Box 3.1).

Box 3.1   The San: hunter–gatherers of the Kalahari

The San people number about 100,000 and occupy semi-arid parts of southern Africa, mostly in Botswana and Namibia. A small proportion of San people still occupy remnants of ancestral lands, mainly in the Kalahari desert (WIMSA, 1998). This minority, especially the !Kung and the G/we people, has attracted considerable attention from anthropologists who consider that they are at least partly representative of an ancient way of living. Our knowledge of the Kalahari's hunter-gatherers owes much to research conducted in the 1950s–70s. The Kalahari 'bushmen' became an object of international curiosity following the publication of Laurens van der Post's (1958) book, *The Lost World of the Kalahari*. In the late 1950s and 1960s a series of anthropologists went to study this 'lost world', most notably Lorna Marshall, Megan Biesele and Richard Lee. This has led to a substantial and detailed body of knowledge about the economics and culture of the !Kung people.

The !Kung traditionally live in small bands of ten to 30 people, erecting temporary villages near to water sources and generally relocating when forced to find a new water source – water is the most scarce resource for humans in this environment. In the wet season, the band can become more dispersed as the need to congregate around water is removed. The forces of production are based on 'stone-age' technology. Women employ digging sticks to find roots, skin bags for carrying plant materials, ostrich shells for carrying water, and wooden, stone and bone tools for food preparation. Men are the main

(continued)

(continued)

hunters in this society and employ bow-and-arrow, poison, spears, snares and sometimes hunting dogs. The women use an extensive knowledge of local plants to provide food, medicines and water (water-bearing roots are an important part of the survival kit). Richard Lee's study revealed that they have names for over 200 plant species and that 105 of these are rated as edible. The mongongo nut is the most abundant food source, yielding its fruits throughout the year and providing a good reason for not adopting agricultural technologies. As one bushman is quoted as saying: 'Why should we plant, when there are so many mongongo nuts in the world' (Lee, 1968, cited in Sahlins, 1972, p. 27).

The !Kung have no formal chief, decisions are made through debate leading to consensus and there are strong cultural norms which favour peaceful resolution over argument and violence. The relations of production are generally described as egalitarian and, despite the strongly gendered division of labour, women enjoy a status which is higher than in most agricultural and industrial societies (Shostak, 1981). The spiritual world is prominent in !Kung society. As is commonly the case, rituals must be performed in order to communicate with the spirit world, in order to protect people's health and to ensure the continued fruitfulness of their environment. In the world of the !Kung, nature is not a simple, external and physical entity. Nature is given meaning through the assignment of human and spiritual significance. To employ a commonly used term for this, 'nature is socially constructed'.

Prior to the 1960s, academics and explorers had largely portrayed the hunter-gatherers' way of life as a fairly miserable existence: being hounded by predators, ravaged by disease and constantly battling against starvation. But this picture has been challenged by Richard Lee, Marshall Sahlins and other anthropologists. The remaining hunters and gatherers, as seen by Westerners over the last 500 years, may well have exhibited some or all of the above problems, but such problems might in large part be attributed to their contact with modern society rather than the intrinsic inadequacy of their mode of production. Hunter-gatherer societies have often suffered the following problems as a result of contact with modern societies:

- the loss of territory to farmers, ranchers and loggers;
- being forced into unfamiliar and inhospitable environments;
- loss of the strongest members to warfare and enslavement;
- being exposed to the ravages of measles, influenza and other diseases to which there is little resistance;
- the introduction of alcohol, leading to the endemic alcoholism which readily afflicts already demoralized people.

Marshall Sahlins' chapter 'The Original Affluent Society' (Sahlins, 1972) provided a compelling rejoinder to the established wisdom that hunter-gatherers typically experienced a poor quality of life. On the contrary, he proposed that those who had escaped the ravages of modern society actually

(continued)

(continued)

experienced a very good life, characterized by affluence. For Sahlins, affluence could be achieved 'either by producing much or desiring little' (1972, p. 1).

Sahlins' work collates evidence that the hunter-gatherer mode of production satisfies modest material needs with a relatively small requirement for work. Indeed, this mode of production is characterized by more leisure time than agricultural or industrial modes of production and, for this reason, Boserup (1965) had previously concluded that a society would only undertake a transformation if compelled to do so by population growth. While Sahlins and Boserup present strong evidence, it should be borne in mind that this apparent affluence somehow spawned cultural practices which seem to reflect material insecurity. Simmons (1997) goes as far as to argue that requesting assistance from the spiritual world reflected a lack of faith in the bounty of the earth.

James Suzman, an anthropologist working with the San in Namibia, is keen to point out that descriptions of the past condition of the !Kung people should not be taken as illustrative of the general condition of any of the San people today (pers. com.). In fact, the work of Richard Lee and his colleagues has itself been the subject of intense debate within anthropological circles in the early 1990s. Revisionists rather harshly contend that it was absurd to think that the bushmen studied were genuinely representative of a pure hunter-gathering tradition.

3.6 Agricultural mode of production

> There is nothing mystifying about agriculture and it is unlikely that any group of people at any time or any place in the world has been unaware of the connection between the seed and the plant . . .' (Clutton-Brock, 1987, p. 47).

If the control of fire was the first revolution in human relationships with the environment, the development of agriculture was the second, establishing unprecedented command over nature. However, if we are to accept the evidence that hunter-gatherer societies commonly enjoyed a good level of subsistence, and that this was gained with only moderate work effort, we must ask why any groups would voluntarily embrace a new technology which was unnecessarily arduous. This is what we might call the **neolithic question**. Having ruled out the idea that agriculture was adopted because it had just been invented (surely an implausible argument), common sense compels a search for some further reason for a Late Pleistocene and early neolithic resource scarcity. While Boserup (1965) asserted that population growth (leading to land scarcity) induces agricultural innovation, this argument may only offer one component of the answer we are seeking. For one thing, it is almost certain that a period of accelerated population growth would have *resulted from*, rather than preceded, this change to the forces of production, in the same way that the next major change to the forces of production (industrialization) was to again

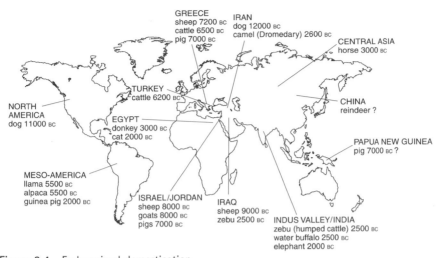

GREECE
sheep 7200 BC
cattle 6500 BC
pig 7000 BC

IRAN
dog 12000 BC
camel (Dromedary) 2600 BC

CENTRAL ASIA
horse 3000 BC

TURKEY
cattle 6200 BC

NORTH
AMERICA
dog 11000 BC

EGYPT
donkey 3000 BC
cat 2000 BC

CHINA
reindeer ?

PAPUA NEW GUINEA
pig 7000 BC ?

MESO-AMERICA
llama 5500 BC
alpaca 5500 BC
guinea pig 2000 BC

ISRAEL/JORDAN
sheep 8000 BC
goats 8000 BC
pigs 7000 BC

IRAQ
sheep 9000 BC
zebu 2500 BC

INDUS VALLEY/INDIA
zebu (humped cattle) 2500 BC
water buffalo 2500 BC
elephant 2000 BC

Figure 3.4 Early animal domestication

Note: A question mark denotes uncertainty associated with insufficient evidence.

Sources: Various; especially Clutton-Brock, 1987.

stimulate accelerated population growth. It seems likely that climate-related environmental change provides the complementary cause of resource scarcity. In particular, increasingly arid climates may have forced population-clustering around remaining water, and stimulated this change in productive technology in eastern and western Asia and the Middle East.

It is also the case that many of the larger mammal and bird species had become extinct through a combination of climate change and hunting, probably making hunting a less fruitful form of labour. In Eurasia, the end of the ice age coincided with the extinction of nine species of large mammal, including the woolly mammoth, bison, woolly rhino and giant Irish elk (Simmons, 1993). However, the decline in availability of large mammals is not in itself a sufficient explanation for the emergence of agriculture. For one thing, the most dramatic Pleistocene extinctions occurred in North America and Australia and these 'new world' regions were relatively late to agriculture.

What we can be more certain about is that, at the end of the ice age, 100 per cent of the world's population followed a predominantly hunter-gatherer mode of production. Various of these groups would have dabbled in forms of proto-agriculture – aiding the distribution of seeds of preferred plant species, using fire to alter the vegetation structure of forests and maybe even diverting streams to irrigate favoured patches of vegetation – but none had domesticated animals or cultivated plants in specially assigned plots of land. Both these facets of the agricultural revolution began 10,000–12,000 years ago, with archaeological evidence revealing the earliest sites in the Middle East (Figures 3.4 and 3.5). By 5,000 years ago, tribal agriculture had become the most widespread mode of production, with pastoral societies (those relying mainly on domesticated livestock), agrarian societies (predominantly plant cultivators) and mixed farming systems (Figure 3.2).

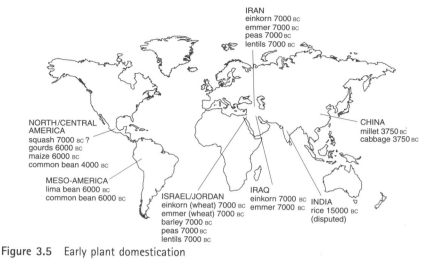

Figure 3.5 Early plant domestication
Sources: Various, especially Goudie, 1990.

Box 3.2　Early agriculture in the British Isles

While the British Isles were visited by migrant populations of hunter-gatherers for several hundred thousand years, permanent occupation probably only began following the retreat of the last ice age. Until 6,000 years ago, hunting and gathering was the sole mode of production, with small groups moving with the seasons and employing flint, bone and wooden tools to obtain and process food. Certain foods such as shellfish and hazelnuts would have been reliably obtained in some areas and such food staple abundance might help to explain why agriculture was not indigenously adopted.

Neolithic people arrived in the British Isles around 6,000 years ago, having gradually spread out from the hearth of agriculture in what is now called the Middle East. The 'neolithic revolution' saw the beginning of wide-scale forest clearance as the oak, elm, lime and other woodlands were cleared through slash-and-burn technology. The shifting cultivation practised was not dissimilar to the better-known tropical form outlined in Figure 3.6. Small plots would be cropped for a few years, after which they would be abandoned and left to regenerate through a succession of vegetation types until either fresh clearance was made or the climax vegetation was re-established. The staple cereal crop planted was 'emmer', a primitive form of wheat that had already been in cultivation in the Middle East for 3,000 years. Barley was later introduced by early bronze age settlers about 4,500 years ago. In 1953 an experiment was conducted to test the ability of stone-age technology to make wholesale clearance of mature woodland. Using original flint axe-heads with new wooden hafts, three men cleared 600 square yards of forest in just four hours (Pennington, 1969). Such an experiment helps us to understand how neolithic farmers could cause large-scale deforestation and have an enduring impact on British landscapes.

(continued)

(continued)

The advent of the plough coincided with the first permanent settlements and the beginnings of communal labour to construct ceremonial sites. Stonehenge was first developed as a place of ceremonial ritual in 2800 BC, with the standing stones erected some 700 years later (Wood, 1978).

3.6.1 Cultivation

Shifting cultivation proved to be a sustainable method of cultivation while population densities remained low. This method of cultivation generally involved the clearance of a small plot of forest or scrub vegetation by *slashing* all the vegetation which existing tools could handle, allowing it to dry for a period and then *burning* the vegetation during a dry period. The burning of vegetation not only clears the land but also releases nutrients into the soil. In tropical forest areas, a typical cycle of cultivation involves cultivation for a period of two–three years followed by a fallow of at least ten years, after which the land might be cleared and farmed once again (Figure 3.6). The reason for abandoning a plot after a small number of seasons is a combination of the depletion of soil nutrients and the vigour of forest regrowth. Soil nutrient depletion leads to falling yields, and the weed problem reduces yields and intensifies labour requirements. Indeed, the weed problem has probably been under-reported as a cause of abandonment. Slashing and burning removes most surface vegetation but leaves tree stumps, live rootstock and buried seeds.

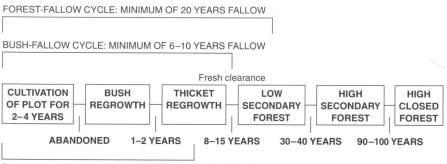

Shorter cultivation cycle caused by shortage of suitable land, resulting in permanent or semi-permanent deforestation and soil degradation

Figure 3.6 Shifting cultivation: generalized cultivation cycles in moist tropical forest

3.7 Tribal relations of production

As you might by now have anticipated, the change to the forces of production known as the 'neolithic revolution', was accompanied by changes to the relations of production and, as will be highlighted later in the chapter, changes in cosmology. Increases in productivity (especially in the quantity of food which each person could produce) led to greater capacity to produce surplus. The

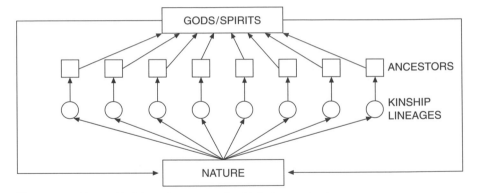

Figure 3.7 Lines of influence in a generalized tribal cosmology
In an extension to hunter-gatherer cosmologies, humans do not take nature's bounty for granted but act to secure abundance, performing rituals that are designed to ensure that the spirits intervene with nature on their behalf. The success of a lineage is therefore dependent on how successful their ancestors are in facilitating this process. Note that some cultures have ancestors who are attached to particular places. For example, among some tribal people of northwest Amazonia, ancestral birth places are used to define the inheritance of kin groups' geographical territories.

potential for surplus accumulation is also greatly increased by the increase in the size of societies, with tribal groups ranging from small units to extended networks of several hundred thousand people. At the same time, the ability to increase personal possessions was facilitated by the sedentary nature of some agrarian societies and by the harnessing of animal transport by pastoral societies. The need for a change in cosmology is evidenced by the fact that many hunter-gatherer bands actually have cultural taboos that specifically forbid crop cultivation: by, for example, forbidding the 'rape of mother earth' (see below). Clearly such taboos needed to be transformed either in advance of, or in response to the need for, the adoption of agriculture. This last statement previews an important question that will be addressed in the next chapter.

Early pastoral and agrarian societies varied greatly in organization but can, for the sake of the brevity of current analysis, be commonly categorized as 'tribal'. This term conveys the idea that a group or clan is governed by internal structures based upon kinship ties and lineages. Kinship lineages are extended families whose economic and social status are not solely determined by lifetime achievements: it is an early system of hereditary class. In addition, each lineage is linked to a pantheon of gods or spirits through its own particular ancestors. Ancestors perform a vital role as mediators, influencing the gods of the spirit world who, in turn, have the power to ensure (or not) that nature treats that particular lineage favourably (Figure 3.7).

The success with which ancestors intervene to secure nature's provision, is ultimately linked in with inequalities of wealth and power. A benign nature brings economic wealth, but it also brings status. Firstly, material wealth is often an uncontroversial sign that the spirits hold your lineage in high esteem. Secondly, wealth can be converted into social status by securing prestigious goods. In Papua New Guinea, for example, tribal groups commonly associate status with the number of pigs owned, while among the Magar of Nepal, ploughed land is more venerable (Lecompte-Tilouine, 1996). However, while one lineage might have a string of successful years, gaining high yields for its labours or not suffering too much from livestock predators, the reality of a tribal system is that existing structures of wealth and power tend to be perpetuated (reproduced) through systems of surplus extraction. One example of a mechanism for such social reproduction, is a system of ritual whereby powerful lineages perform 'favours' for less powerful ones (such as presenting them 'gifts' or loans, or communicating with important spirits on their behalf). Rules regarding the paying back of such favours can lead to an ongoing transfer of wealth (Peet, 1991). The most powerful tribal members frequently take on supernatural powers, becoming shamans who mediate between earthly and spiritual domains. The performance of such vital roles serves to enforce the earthly hierarchy of lineages, rendering it far more persistent than a system entirely based upon material well-being. The reliance on the spiritual powers of dominant lineages also provides the ideological justification for reproducing such relations of production: a change to this system of surplus extraction would threaten an order which maintains good relationships with the spirit world and with nature.

3.8 Tribal cosmologies

The term **cosmology** refers to the aggregate bundle of cultural endowments which frame the way in which we perceive and act in the world around us. It is a culturally acquired set of beliefs which results in deep-rooted assumptions about the physical operations of the universe, the role of spiritual dimensions, what is right and wrong, what is beautiful, our relationship with nature, and so on. For example, in twentieth-century 'Western' culture, people tend to hold some (but not necessarily all) of the following beliefs:

- A single God exists.
- Time is linear (rather than cyclical).
- Humans are different to animals.
- Humans evolved from apes.
- The human mind contains a subconscious.
- Mountains are beautiful.

Such constituents are said to be culturally acquired because they are not universally held truths. They are part of the cosmology of certain people, in certain places, in certain historical periods. The cultural endowment 'mountains are beautiful' provides an interesting example when considered in relation to the UK or the United States. That this is a part of our recent cultural heritage is little doubted – for evidence, we need look no further than the fact than mountainous areas have dominated the selection of National Parks, supposedly our most valued rural landscapes. But the unconditional beauty of mountains has not always been a component of our perception of nature.

Mountains have historically been fearful, ugly and untamed places, yet to benefit from the beautifying process of human domination (Tuan, 1974; Short, 1992).

3.8.1 The vertical cosmos

Tribal and hunter-gatherer cosmologies are vertical, stratified and hierarchical. Typically, the world is viewed as being composed of three interrelated spheres of existence (the ontological framework). These are the sky (stars, heavens or sometimes mountain tops), the earth and the underworld. This vertical strati-fication is often further divided at a more localized level, with some tribal people perceiving the sky as itself being made up of a series of hierarchical layers. The basic framework of a vertical, three-tier universe is widely distributed among existing or near-recent tribal people in South and Central Asia, South America and Sub-Saharan Africa. For the Magar of Nepal, for example, the world is composed of three levels. The celestial sphere is made up of a small number of gods who reside in mountain tops. These higher gods govern the whole earth and have the power to control 'natural' forces such as rain and disease. Reproduction of social stratification is linked to these gods in a very tangible way: people of lower-caste groups, and women in general, are not allowed to communicate with these supreme deities directly and must there-fore rely on the favours of higher-caste groups. The earth is the second level of the cosmos. It is cohabited by humans and petty deities and there is a connection to the third level, an underworld, through the forests. Forests include an underworld inhabited by deities such as the earth goddess 'Bhume' (Lecompte-Tilouine, 1996).

3.8.2 Animism and totemism

Animism and totemism are concepts that describe two different, but related, ways in which premodern human societies construct their relationship with non-human nature. Totemic systems construct an understanding of human society which is built upon references to animals, plants and other natural entities. Clans, kinship groups or individuals are related to particular natural phenomena through ancestral lineage and are symbolically associated to that part of nature. Such relationships are typically held with one or more animal species (such as a wolf or an eagle), but may be with other objects. Recent anthropological studies suggest that the relationship is used as a way of ex-plaining behavioural differences: those associated with the jaguar have 'jaguar-like' qualities, those associated with the raven have 'raven-like' qualities and so on (see e.g. Arhem, 1996). However, such a strong behavioural association is certainly not universal. For example, when Northwest Coast American Indians claim an association with a killer whale, they are not saying that they behave like one; rather, they are expressing their belonging to a kinship line-age which descends from an ancestor who had a special relationship with the killer whale (Halpin, 1981). In other words, the association is a way of under-standing kinship links rather than individual behaviour.

Figure 3.8 The Warlis are a tribal group of shifting cultivators, living in the Western Ghat hills of Maharashtra, Western India. The close relationship between their daily life and nature is depicted in much of their art
Source: Adrian Martin.

As well as using the natural world as a model for establishing difference between individuals and between kinship groups, it can also be used for understanding the basis of social stratification. A totemic system is hierarchical because certain species are more prestigious than others. The most important kinship lineages descend from ancestors who had close relationships with particularly prestigious species and, in the case of Northwest American Indians, displayed their family status through the depiction of animal 'crests' on totem poles. As such, the natural order serves as a blueprint for understanding human social stratification.

Animism can be thought of as the reverse of totemism, although this does not mean that the two systems are incompatible. Whereas totemic systems model social relations on nature, animic systems model nature on social relations (Arhem, 1996). Here, human characteristics, such as the possession of a soul, are endowed upon animals, plants and even non-living objects such as stones. This practice is common among American Indians, Australian Aborigines and in Asia. In contemporary South Asia, one can see animic systems of belief crystallized within Hindu lore, with species such as elephants, cows, cobras and monkeys strongly associated with the Hindu pantheon of gods. In some tribal societies in Amazonia, animals are thought to be humans in disguise (Arhem, 1996; Clastres, 1998). Societies with animic cosmologies construct nature (assign meaning to it) by filling it with human qualities.

To characterize totemism and animism as fundamentally different is to overlook a crucial commonality. Both systems of belief express considerable continuity between society and nature, a cosmology which is to be contrasted with the radical separation of society and nature to be found in modern industrial society. Combined within a single cosmology, animism and totemism form an interesting co-evolutionary outlook on nature's relationship with society. In such a cosmology, society and nature mutually constitute each other and it is impossible to understand one without reference to the other.

3.8.3 Mother Earth

One of the most widespread cultural expressions of animism involves understanding the Earth my imbuing 'her' with the qualities of a human mother. Anthropological evidence points to the enormous importance of this component of people's cosmology, particularly as a means to reproducing existing forces of production by building a system of moral values (part of an ideology) which precludes activities such as ploughing and mining, which are harmful to the 'mother'. Verrier Elwin (1939) describes the significance of *bewar* to the Baiga tribe of central India. *Bewar* is the Baiga's name for their particular form of shifting cultivation which follows the kind of cycles identified earlier in the chapter but which involves no ploughing of the earth: seeds are planted in the layer of ash left after burning, rather than in the underlying soil. Such a practice is reproduced through an oral tradition of mythology in which very clear instructions are passed down from the Bhagavan: 'You must not tear the breasts of your Mother the Earth with the plough. . . . You will cut down trees and burn them and sow your seed in the ashes' (p. 250). Such wisdom is not take lightly and Elwin reproduces the plight of a man who confessed to having used the plough: 'But my children have always been weak and sickly on account of it. If even one Baiga in a village touches the plough, we are all affected' (p. 250). Elwin notices the similarity with American Indian cultures, citing the 1870 pronouncements of a prophet from the Colombia River basin: 'You ask me to plough the ground. Shall I take a knife and tear my mother's bosom?' Similar beliefs can be maintained within some societies who actually live by the plough. The Brahmin caste of the Gulmi people of Nepal, for example, will not engage in agriculture because they do not want to 'wound their mother' (Lecompte-Tilouine, 1996). In this case, however, it is acceptable for lower-caste groups to plough the earth on their behalf.

3.8.4 Earth goddesses

Among tribal people, earth goddesses often play a crucial part in ritual life despite the fact that they are not the most elevated of spirits within the overall pantheon of gods. Earth goddesses, in various guises, tend to inhabit the earth or the underworld rather than the divine celestial sphere. They tend not to be entirely benevolent and are variously referred to as wild and untamed, evil and dangerous. For example, Xama, the earth goddess of the Hani people of China, is considered to be a 'monstrous cannibal' (Bouchery, 1996, p. 103). However, while such goddesses are potentially dangerous, societies generally manage to hold a form of pact (first initiated by the ancestors) whereby their ritual devotion to the goddess is repaid with her protection. This takes the form of secured livelihood and a correctly ordered relationship with the natural environment (Formoso, 1996).

The need to exercise extreme caution when dealing with nature raises questions about the environmental credentials of premodern people. For example, did their respect for nature, their apparent concerns about scarcity, and their associated restraints on the use of nature constitute early systems of environmental management? Jumping to conclusions on this question is not advisable, for it is a contentious issue requiring careful consideration. Such a consideration is reserved for the next chapter.

3.9 Conclusion: adaptation or construction?

This exploration of the changing relationship between societies and their environments in the premodern world has identified two important processes which do not sit comfortably together and which therefore require some resolution. On the one hand, it can be seen that human societies adapt to their environment, evolving forces of production which are appropriate to the available resources and, correspondingly, developing a cosmology which makes sense of this relationship with nature. Neolithic people, for example, had to adapt to a changing environment through the adoption of agricultural technologies.

On the other hand, however, societies are also seen as constructing their environments, both in terms of physically modifying them and in terms of filling them with human constituted meaning. The physical process is exemplified through the use of fire to modify local environments, the breeding of domesticates and the farming of plant crops. The imposition of meaning is exemplified through animistic systems of belief.

These two observed processes (adaptation and construction) have tended to be viewed as contradictory – you can believe in one or the other, but not both of them together. In other words, you either believe that the evolution of human culture is a direct response to the need to survive within a particular environment or, alternatively, you believe that human culture evolves relatively independently and is able to impose itself onto a passive environment. Two schools of thought have therefore developed, with the early part of the twentieth century seeing the 'cultural adaptation' school gain prominence and the period since witnessing the rise to supremacy of the 'cultural construction' school. Most recently, there has been a move towards resolving the (apparent) contradiction rather than crudely plumping for one explanation over the other.

The starting point for a resolution is actually to question the fundamental assumptions utilized in modern enquiry about human–environment relations.

Chapter 2 explained that modernity has involved a radical separation of nature and society, in which nature is seen as external to us, rather than a part of us. To a critical realist, it is this assumed structural relationship which is erroneous and which has led to the contradiction in question. For if you transcend this separation and view society and nature as structurally united (as it happens, premodern people did just this) the issue of which determines the characteristics of which becomes blurred. In such a 'dialectical' analysis, one can only talk of social and natural processes mutually constituting each other (a phrase also introduced in Chapter 2). Both construction and adaptation become part of a dynamic, complex and co-evolutionary process of change which is acted out through humans functioning within nature, i.e through the use of nature in the processes of production.

3.10 Further reading

The classic geography textbook is Simmons (1996), 2nd edn, *Changing the Face of the Earth*. For more detailed consideration of the concept 'mode of production' and its application to societal change, Richard Peet's (1991, new edition forthcoming) *Global Capitalism: Theories of Societal Development* will serve you well.

Within anthropology, you should certainly review Marshall Sahlins (1972) 'Original Affluent Society', in *Stone Age Economics* or (in abridged version) in Rahnema and Bawtree (1997) *The Post-Development Reader*. Also browse Lee and Daly (1999) *The Cambridge Encyclopædia of Hunters and Gatherers*. For reading on tribal agricultural societies, try Verier Elwin's (1958) hugely entertaining *Leaves from the Jungle* or Clastres' (1998) *Chronicle of the Guayaki Indians* that provides a rich case-study of a tribal group in Paraguay.

Chapter 4

Premodern state societies

4.1 Introduction

By 5000 BC, the neolithic 'revolution' was well advanced and this transition from hunter-gatherer to agricultural societies had created the conditions under which urban settlements first appeared in Mesopotamia, Egypt and the Indus Valley. While the dawn of urban settlement involved a change in the organized use of nature, it was also accompanied by more centralized systems of social organization based upon religious and monarchical authority. Premodern state societies span a period from 3000 BC to the near present, and display a wide variety of social and environmental relations. This chapter begins by briefly examining the intensification of environmental transformations that began with the urban revolution. On the one hand it explores the evolving patterns of human transformation of nature and, on the other hand, it examines the related changes in social organization. Having established some general features of premodern states, a more detailed study is made of late medieval England.

4.2 The urban revolution and the birth of the state

Hunter-gatherer and tribal agricultural societies developed localized forms of governance, and it might be argued that these early decision-making procedures provide the first examples of political activity. Nevertheless, such societies can be accurately described as 'stateless' because they did not contain any of the features and institutions of **state society**: clearly defined territories, a legal system, armed forces and centralized authority. In the absence of a state, the reproduction of a particular economic relationship with nature depends upon its ecological sustainability (e.g. continued supply of animals to hunt) and its ideological sustainability (i.e. whether the evolving system of beliefs and values continues to sanction the existing forces and relations of production). With the emergence of state institutions, we have to consider a further factor in the reproduction and transformation of society, as indicated in Figure 3.1. The state can be a force of change but its primary function is to maintain an existing mode of production: through the use of legitimized force to sustain law and order, through the use of education to sustain the dominant ideology and through the protection and provision of human and ecological capital.

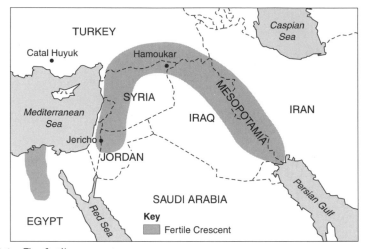

Figure 4.1   The fertile crescent
Source: Based on Morris, 1994; Mellaart, 1975.

The word 'civilization' is derived from the Latin for a city. While the word now has rather more complex connotations, the formation of urban centres can still be considered as the beginnings of civilization and state society. The first cities developed in southern Mesopotamia and Egypt, in low-lying areas on the fertile floodplains of the Tigris, Euphrates and Nile Rivers (Figure 4.1). This **urban revolution** represented the culmination of a millennium of rapid technical advance from 4000 to 3000 BC, which included the development of writing, the spread of wheeled transport and sailing ships, the invention of bronze, methods of harnessing flood waters for irrigation and large-scale drainage systems. The location of these first cities demonstrated that early Sumerians (the people of Sumeria in southern Mesopotamia) and Egyptians had success-fully created agricultural environments on fertile soils with inadequate rainfall and drainage, what Simmons (1989, p. 93) describes as 'one of the most out-standing environmental transformations in ancient times'. Indeed, ecological capital was rather poor in Sumeria: there was a lack of stone and timber for construction, lack of minerals and the climate was arid (Morris, 1994). Con-trol over river waters was essential to an intensification of population densities and it was also an important source of political power. For example, the Sumerian city of Kish occupied a critical location on the Euphrates, thus controlling water to a wide network of irrigation channels (Heise, 1996). In cities such as Kish and Ur, the environmental project was inseparable from the sociopolitical project and it may be no coincidence that they both enjoyed periods of pre-eminence over Sumeria's other city-states.

The development of bronze tools contributed to the first physical charac-teristic of cities: the construction of large 'monumental' buildings, including temples and palaces. Bronze also led to the development of superior weaponry which necessitated the second physical feature of cities: the construction of fortifications. Temples and palaces drew people in during times of peace and fortifications kept others out in times of war (Hallo and Simpson, 1971). Early

---

Box 4.1   Archaeological uncertainty and controversy

Çatal Hüyük in Anatolia and Jericho in Jordan are large settlements that significantly predate the earliest Sumerian cities. Drawing on the discovery that Çatal Hüyük may be as early as 8000 BC, Jane Jacobs challenges the conventional wisdom that agricultural development is a vital prerequisite for city formation (Morris, 1994):

> For this purpose Mrs Jacobs invented the imaginary city of New Obsidian, which was the centre of a large trade in obsidian, the tough black natural glass produced by some volcanoes. . . . In 8500 BC New Obsidian's population numbered about 2,000 persons, a large proportion of whose food 'is imported from foreign hunting territories. This food, which consists overwhelmingly of live animals and hard seeds, is traded at the barter square for obsidian and for other exports of the city' (Morris, 1994, p. 18).

So Jacobs envisioned the possibility of a city that feeds itself through trade with hunter-gatherers, rather than through an agricultural hinterland. Morris rejects the plausibility of such a pre-agricultural city economy:

> A hunting-gathering catchment area of at least 20,000 square miles would have been needed for an urban population of 2,000 persons, giving a circle of radius of about 80 miles. How could the food supply have been taken into the city, in quantity, over such considerable distances, when wheeled transport was still 5,000 years in the future? Or stored, without pottery technology? (Ibid., p. 18).

Both Çatal Hüyük and Jericho are now considered to have been large villages with proto-urban characteristics; they do not therefore question the established orthodoxy that the oldest cities were Sumerian, dating from around 3500 BC. However, Howe (2000) documents the recent discovery of Hamoukar, located in present-day Syria, between the Euphrates and the Tigris. Early excavations are suggestive of a walled settlement of about 30 acres, with large communal ovens and artifacts that point to a division of labour akin to urban society. Most significantly, this north Syrian society appears to have been already flourishing by 4000 BC. The big question is whether or not this was an urban civilization that predates Sumeria.

---

cities are defined by function rather than size: they are centres of trade, culture, religion and government. In addition, they might also be considered as centres of agriculture, due to the close connection between the economies of city and countryside and due to the control which the city elite held over surrounding agricultural areas. The development of early cities is not thought to have involved rural depopulation; rather, urban centres led to higher population densities in the vicinity, resulting from the city's requirement for surplus food production and the resultant concentration of agricultural production. The construction of 'new', urban environments was allied to a reconstruction of agrarian environments (Box 4.1). In the case of Mesopotamia, this rural reconstruction was ultimately to prove unsustainable: the combination of irrigation and inadequate drainage led to serious salinization of soils (see Goudie,

Figure 4.2 Excavations at Hamoukar. Mud brick architecture and pits of 4th Millennium BC; the partially excavated circular feature in the foreground is a large, communal oven, an important clue to a division and specialization of labour. Photograph courtesy of the Oriental Institute, University of Chicago © 2000.

0   1   2 cm

Figure 4.3 Stamp seal from Hamoukar, 4th Millennium BC. The seal is made of bone in the shape of a lying deer and the reverse side depicts three horned animals, possibly ibexes. Seals were used to make imprints in clay, to seal containers of produce, to keep account of origins and to make it clear where something had been opened. They were an important means of administering transactions and a precursor to written records. Photograph courtesy of the Oriental Institute, University of Chicago © 2000.

1990; Hughes, 1994), a problem which is still familiar in semi-arid modern farming environments in Australia and elsewhere. Today, one-quarter of Iraq's land is thought to be saline (Morris, 1994).

4.3 Tribute and accumulation

The transition to premodern states involved a change in political affinities. Kinship loyalties were replaced by the spiritual and military protection afforded by priests, chieftains and kings. Ancient Sumeria was governed by a collection of kings, each with control over a city-state, with one having supremacy at any one time; the Chinese Empire (200 BC–AD 1912) was ruled by a succession of dynastic rulers, as was Egypt (3200–525 BC) and Babylonia (1792–539 BC); the Mayan civilization (1500 BC–AD 900) was ruled by a class of warrior-priests, as were the other main meso-American civilizations of the Aztecs (AD 1400–1521) and the Incas of Peru (AD 1440–1532). While military strength became an important source of power in many early civilizations, it must be remembered that the secular state is a feature of the modern world and that the supernatural basis to power found in tribal societies was far from diminished. Class continued to be related to access to the spiritual world and the associated ability to intervene in the workings of the natural world. For example, the religious leaders of the Maya, Aztec and Inca civilizations derived legitimacy from their shamanic powers: the ability to move between the earthly and spiritual domains, interpreting and influencing the intentions of deities. The legitimacy of premodern rulers was generally rooted in a spiritual hierarchy.

The evolving spatial configurations of production went hand-in-hand with evolving sociopolitical institutions. Increasingly productive agriculture provided the condition under which cities and new social orders could persist. As the productive use of nature became associated with capital accumulation, there was often a move by rulers to control the sources of production (natural capital), a process whereby nature was constructed as a source of political power. In premodern states, agricultural surplus is appropriated by dominant classes of priests and rulers, in the form of a tribute or tax. Typically, surplus is extracted in the form of a share of the crop or in the form of labour. *Corveé* labour is a form of tribute in which an amount of unpaid work, say one day per week, is demanded by the ruler. At the extreme end of unfreedom, all rights to an individual's labour could be owned by another individual – the system of slavery prevalent in Babylonian, Egyptian, Greek and other premodern (and modern) civilizations. Accumulated capital was realized in the amassing of personal possessions, the waging of wars and also in monumental constructions such as the palaces of Sumeria and Babylonia, the pyramids of the Egyptians, Mayas and Aztecs and the Great Wall of China (completed during the Qin dynasty, 221–206 BC). The accumulation of capital also encouraged a division of labour which went beyond that found in tribal societies. In addition to an increasingly gendered division of labour, labour became more specialized and new 'urban' professions emerged, including soldiers, bureaucrats, artisans, entertainers and intellectuals.

Box 4.2   Wittfogel's hydraulic society

Societies based on a tributary mode of production vary significantly in the level of state control over the environment and in the level of surplus appropriated by ruling classes. While feudal Europe can be considered as a weak, devolved form of tributary society, ancient Egypt, Mesopotamia, India, China and others displayed much stronger state control, exemplified by more rigorous and centralized extraction of surplus. Karl Wittfogel used his study of Chinese society to derive an interesting theory which sought to explain such variations. He investigated the geographical distribution of 'despotic' relations of production, ones where an exploitative system of surplus extraction was maintained for long periods of time. The ambitious nature of the thesis resulted from the fact that he sought to explain both the causes of the origins of such states and the reason for their stability. Wittfogel (1957) observed that despotic societies appeared in regions with insufficient rainfall for non-irrigated agriculture. This observed relationship led him to believe that environmental conditions provided an important determinant of the kind of state society which was likely to emerge. More specifically, he viewed the despotic social order as a necessary cultural adaptation to the environmental condition of water scarcity. The need for widespread systems of irrigation necessitated the existence of a powerful state which could appropriate surplus products and labour and utilize them for the construction and maintenance of large-scale irrigation works. By the same measure, the subsequent stability of Wittfogel's 'hydraulic societies' resulted from the people's dependence on the state to maintain such provision. The ideology which supported the despotic relations of production was therefore founded on people's belief in the essential nature of the state's irrigation work.

Wittfogel is generally correct in linking strong tributary societies with semi-arid environments and, correspondingly, in linking less-developed tributary systems such as feudalism with temperate climates. However, his work has been criticized for a number of weaknesses (O'Leary, 1989; Hindess and Hirst, 1975; Atkins, Simmons and Roberts, 1998). The main analytical flaw arises from his determination to explain the origins of the despotic state. In this department, the thesis simply doesn't add up: if large-scale irrigation is a *consequence* of a strong state it cannot also be the *cause* of the state. In fact, many of the states in question developed prior to any extensive, large-scale irrigation system. Nevertheless, Peet (1991) is justified in seeking to salvage something from Wittfogel's work. While it can't explain the development of the despotic state, it does provide an argument which links the state's control over nature with its own stability. Likewise, Harvey (1996, p. 184) supports the view that 'one path towards consolidation of a particular set of social relations . . . is to undertake an ecological transformation which requires the reproduction of those social relations in order to sustain it'. Irrigation works on the Narmada River in India provide an excellent contemporary example. Arundhati Roy (1999) skilfully describes this project as an assault on freedom, enforcing a terrible and irreversible dependence on the state.

Figure 4.4 European overseeing irrigation workers, Madras, India. Unknown Indian artist, 1785
Source: National Arts slide Library Record, © De Montfort University.

4.4 Feudal England: AD 1066–1381

Feudalism can be considered as an ideological and an ecological project in which the institutions and hierarchies of the social order are expressed in the spatial organization of environments. As Dodgshon (1990) puts it: 'feudal relationships were mapped onto the landscape'. It might be argued that this relationship between the social and ecological orders invested feudalism with a source of strength and stability. However, in contrast to Wittfogel's analysis of semi-arid environments (Box 4.2), it might also be argued that it contributed to its eventual demise, an assertion that will be developed below.

The use of the term 'feudalism' is currently under review in historical circles and it cannot be employed without some opening reservations. Both the duration and spatial extent of feudal relationships are under scrutiny and the following three points should be noted. Firstly, Reynolds (1994) argues that the practice of assigning feudal relations to ninth- and tenth-century England is erroneous. Accepting this argument, we consider English feudalism to begin with the Frankish system imposed by William the Conqueror. Secondly, even within this late medieval period, feudal relationships were not fully and evenly expressed throughout the land. As with hunter-gatherer and tribal modes of production, a generalized set of relationships expressed itself in difference and diversity as much as it expressed itself in conformity. Thirdly, the common understanding of feudalism as a relationship between lord and peasant must be incorporated into historians' more technical use of the term to describe a system of relations among members of the ruling military.

Feudalism can be understood in terms of four sets of relationships.

1. Relationships among the military and religious elite
Feudalism can be considered as a weak form of tributary governance because its very *raison d'être* is a weak monarchy which can only survive through devolution of power. In principle, the divine right of feudal kings granted them absolute power and ownership over all peoples and lands. In practice, both land and power were devolved to a hierarchy of vassals, an elite who were granted parcels of land (fiefs), in return for military service. The greater vassals, with extensive fiefdoms, might in turn devolve smaller fiefs to lesser nobles. Vassals could live off the labour of peasants (see below) and in return had to go to war for the king or, as became increasingly the case, pay 'scutage', a sum of money which enabled the king to hire mercenaries instead.

2. Relationship between vassals and peasants
Vassals exercised lordly powers over both land and people within their fiefs. The system most properly known as 'manoralism' involved the division of land into 'demesne' land and other land which was leased out to tenant farmers. In the most highly feudalized manors, the majority of peasants became bonded tenants (villeins), forced to give up many significant freedoms in return for the right to farm their tenancy. In addition to paying rent, fines, marriage and death duties, taxes and tithes, they were also expected to give part of their labour to the working of the vassal's demesne lands. Villeins were not free to leave the manor and if they attempted to do so, the vassal could send bailiffs to force them to return (see e.g. Britnell, 1990). In addition, villeins were deprived of common law rights in respect of their lordly vassal and were therefore highly vulnerable to exploitation. An additional class of free peasant secured tenancy through the payment of a higher rent, offset by fewer obligations to the vassal.

3. Customary relationships
Feudalism, in common with any emergent social system, was not imposed onto blank slates. Places are endowed with an historical momentum and new feudal relationships were superimposed onto pre-existing customary rights

which had evolved over the centuries. Thus the tyranny of feudal relations was often mediated by less draconian customs. The fact that feudal relations (the general) were expressed and mediated through custom (the local) contributed to the mosaic of difference which evolved: history merged with geography.

4. The relationship between social and ecological organization
Political power and the accumulation of capital were sustained through the particular relationships between people and environments, especially through the organization of land use. The bonding of peasants overcame the problem of labour shortage and facilitated the flourishing of the economy in the twelfth and thirteenth centuries. However, Postan (1972) argued that this relationship ultimately constrained technological change and that prevailing relations became pressured by an ecological as well as a social crisis.

4.5 Environments in feudal England

Much of lowland England was dominated by the open field system of agriculture, considered to have its origins at least as far back as the seventh century. Typically, two or three large open fields were farmed communally, with each household having tenancy of a number of strips of land which were often dispersed around the fields. Dispersion of plots led to the formation of nuclear villages. With a single consolidated holding, it makes sense to locate the homestead adjacent to the land, leading to dispersed settlements; with dispersed land holdings, it makes more sense to settle in a central location. In a three-field system, two would be farmed while one lay fallow, with annual rotation of use. In addition to the arable fields, livestock usually played a vital role in the farming system. Meadow, pasture and forest lands provided vital grazing for sheep, cattle and pigs, and the livestock provided manure for the fields. Oxen were the favoured draught animals, with horses (which are faster) gradually gaining favour throughout the period. Farming was entirely organic with soil conditioning mainly restricted to animal and human manure, lime, leaf mould, wood ash and, in coastal locations, seaweed. While wheat, rye, barley, oats and legumes were the principal crops, manorial systems often contained a wide variety of produce (Figure 4.5).

The twelfth and thirteenth centuries witnessed a period of economic expansion, characterized by population growth, colonization of new lands by agriculture, urban growth, an advancing material culture, increased domestic and international trade and the commercialization of production. Agricultural production grew to meet rising demand through a combination of intensification and extension, although productivity was also aided by the warm epoch which lasted from the eleventh to the thirteenth century. Intensification relied mainly on improved husbandry techniques such as better weed control and, with the exception of the spread of watermills, windmills and horses, it was a period of relatively minor advances in labour-saving technologies. Extension of agriculture involved the colonization of areas with heavy clay soils, forest lands and wetlands. Improved iron tools enabled the working of clay soils while

Source	Produce	Source	Produce
Arable crops		**Orchard and garden crops**	
wheat (spelt, club, bread)	bread, ale	apples	fruit, cider
oats (cultivated and wild)	bread, pottage, livestock-feed, ale	pears, cherries, figs, walnuts, damsons, plums	fruit and nuts
barley (hulled, naked)	ale, bread, livestock-feed	vines	wine
rye	bread	flax	linen
peas, beans, vetches	whole plant for human and livestock-feed	hemp	rope and linen
		herbs	seasoning, medicines, dyes
all cereal straw	livestock-feed, thatching	leeks, onions, borage, mustard, peas, beans	vegetable foods
Livestock		**Natural resources**	
pigs	meat	deer	meat, manures
sheep, goats	wool, milk, manures, meat, skin for parchment	wild boar	meat
		birds	meat
cattle	draught power, milk, cheese, butter, curds, meat, leather, horn	fish – from fish-pond, river, sea	meat
		hares	meat, fur
horses	draught power, leather	oak and beech trees	acorns and mast for pigs, timber
		other trees and shrubs	nuts, berries, fruits, timber, browse, fuelwood
poultry (chickens, geese, swans, peacocks)	eggs, meat	ferns, bracken, sedges	thatch, bedding, litter
		nettles	linen
pigeons and doves	meat, manures	osiers, reeds	baskets, fish traps
bees	honey, wax	holly, thorns	threshing flails
rabbits	meat, fur	peat	fuel
		herbs	medicines, vegetables
		grass	hay
		grass turves	roofing, fuel

Figure 4.5   Produce from a typical manorial estate
Source: Pretty, 1990, p. 4.

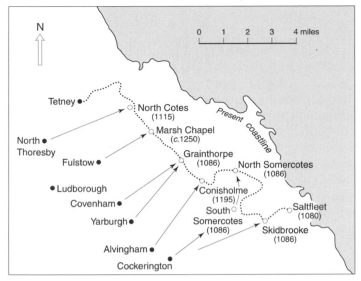

Figure 4.6 The early colonization of English marshlands
The early Anglo-Saxon villages (and North Thoresby, which is Danish in origin) are marked thus:●. The probable course of the Sea Dike constructed by these villages before the Norman Conquest is indicated by a broken line. Daughter-villages founded on the Dike are shown thus:○ and the earliest reference to them is indicated by a date in brackets. The marshland between the parent villages and the Dike had been reclaimed from the sea by the eleventh century for the most part; and the earliest settlements on the Dike were probably of huts occupied in the summer months only (as is suggested by the name Somercotes, 'summer cottages or huts'. Beyond the Dike lie no villages but only single isolated farms of varying dates from medieval times onwards. The parent villages were all founded between 20 and 40 feet above sea-level.
Source: Hoskins, 1955, p. 80.

the construction of drainage channels allowed wetlands such as the Somerset levels and Romney Marsh to be used as pasture. Perhaps the most spectacular environmental modifications occurred along the south and east coasts, where the fenlands were drained and where coastal land was reclaimed through banks, dikes and sluices (Reed, 1990). Large new agricultural environments were developed in Lincolnshire (see Figure 4.6) and Norfolk, contributing to the relative prosperity of these counties. In the north of England, the population had been devastated by the Normans in the late eleventh century. Establishing manors enabled surviving populations to be clustered and farming to be organized on empty lands (Dodgshon and Butlin, 1990).

4.6 Forests

By the time of the Norman Conquest the English landscape and economy were dominated by pastoral and arable agriculture. Celtic, Roman and Anglo-Saxon people had extensively cleared forest and no more than 15 per cent of England remained wooded (Schama, 1995). While thick oak, ash and beech

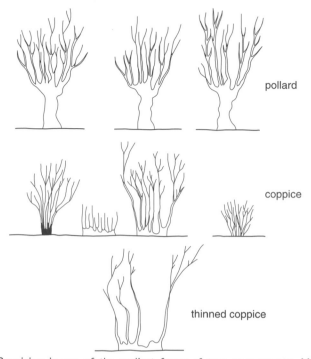

pollard

coppice

thinned coppice

Figure 4.7 Coppicing is one of the earliest forms of tree management. Many trees will coppice, i.e. throw up new shoots from the stool; stumps are cut off near the base to provide small-diameter wood for domestic purposes. During the growth of each cycle, some shoots can be cut small while others are left to grow larger. Pollarding was a later development; new shoots are kept out of the reach of browsing domestic animals
Source: Westoby, 1989, p. 34.

woodland was still present in many parts, it was not wild, untamed nature. Forests were mainly managed for agricultural inputs and for timber production. Timber remained the main source of household fuel and, in the form of charcoal, the main fuel for iron smelting. It was also the main construction material, used to make the peasant's 'longhouse', roofing for stone buildings, agricultural instruments and ships. Woodland was generally coppiced, providing a regular harvest of a suitable range of timber sizes. Cattle, horses, sheep and goats grazed in forests, while pigs – a key source of livelihood in the woodland economy – fed on the glut of acorns and beech mast available in the autumn months. Additionally, fruits, nuts and honey were collected for household consumption and animals were hunted for food and hide.

It is against this backdrop of a functioning woodland society that one has to assess the alienation from forest resources imposed by feudal rule. Under the Normans, 'forest' was a designation which referred to an administrative regime as much as to a vegetation type. This designation could be assigned to unwooded lands, and 'forest' land had been extended to between one-quarter and one-third of the geographical area of England by the end of the twelfth century. These 'Royal Forests' were the preserve of recreational hunting for

Figure 4.8 Buckingham thick copse, part of the ancient Forest of Whittlewood, UK. The copse was managed as 'coppice-with-standards' from medieval times until the late nineteenth century. This photo shows pollarded oak trunk (left), coppiced hazel (centre) and oak standard (right)
Source: Adrian Martin.

the aristocracy and their protection was often rigorous, with peasants stripped of many customary rights of access. Hunting of deer was forbidden, arable land could not be fenced off to protect it from deer, and pigs were excluded, except during autumn months (Reed, 1990). Punishment for real and alleged transgressions of forest rules could be harsh, and this in part reflected the significance of the hunt among royal circles: 'The hunt was not merely a kill that gave potency and authority to the aura of the royal warlord, it was also a ritual demonstration of the discipline and order of his court. No wonder it became a form of treason to spoil the king's aim' (Schama, 1995, p. 145).

The 'Charta de Foresta' of 1217 (two years after Magna Carta) relinquished some of the sovereign power over forests and also reduced the extent of the Royal Forest designation. This forest charter was not the result of peasant disquiet (although this certainly existed), but stemmed from the opposition of the church and lesser nobles and, perhaps above all, the ever pressing need to fund expensive foreign warfare. Forest rights were sold off, enabling clearance for agriculture, timber-extraction and charcoal-burning. In 1204, for example, King John sold the nobles of Devon the right to disafforest the entire county, barring Exmoor and Dartmoor (Hoskins, 1955; Reed, 1990). The social implications of removing 'forest' designation are somewhat unclear. On the one hand, peasants renting land from these new forest-owners were generally freeholders, paying rent on a mostly commercial basis, with few of the bonds of villeinage. Such relative freedom meant that these environments, along with other lightly feudalized environments, became dynamic areas of production, with cottage industries increasingly connected to markets. On the other hand, the new landowners could seek to retrieve their outlay through even more draconian policing and the imposition of hefty fines. The various legends of Robin Hood, widespread by the fourteenth century, pay credence to this, portraying a peasant backlash against the landowning nobles, bishops and monks, and the petty tyrants who administered their rule (Keen, 1991). Yet it is the manner of administration, rather than the social system itself which was the object of their violence. The king's right to own and rule the land remained unquestioned (at least within the Robin Hood legend). Indeed, one of the main sins of the forest (and manorial) tyrants was that their despotism was a form of disloyalty to the king.

4.7 The decline of feudalism: an ecological crisis?

The decline of feudalism in the fourteenth century can be attributed to a combination of events and processes: the centralization of power, increased commercialization and trade, the crumbling bond between king and warlords, the stifling of technological progress, falling agricultural productivity, a steady increase in free tenancies and growing peasant unrest. Postan (1972) emphasized the ecological basis of feudal unsustainability, a thesis which has attracted considerable attention since. Between 1086 and 1300, the population of England roughly tripled from two to six million people, fed largely by the expansion of agricultural production. For Postan, the limits to growth had been reached and, in the absence of technical solutions, peasants became increasingly impoverished through environmental constraints. The conversion of meadow and pasture to arable land had led to a lack of draught power and manure; the lack of manure led to a decline in soil fertility and reduced crop yields; newly colonized lands were often marginal for agriculture and early success was followed by rapid reductions in fertility; sustaining agricultural yields became dependent on rising labour inputs, leading to a fall in the ratio of output to input and a consequent decline in wages; inheritance led to the fragmentation and diminution of individual landholdings; food scarcity resulted in high prices, creating an obstacle to urban growth.

The broad scenario of such a crisis can be supported, although the evidence is not consistent. In Winchester, for example, agricultural productivity was low, but the reason may have been isolation from markets rather than falling soil fertility (Dodgshon and Butlin, 1990). Certainly, the growing urban sector meant that accessible markets acted to stimulate productivity in some areas, and location (as opposed to environmental conditions) became an increasingly important determinant of improved husbandry. Pretty (1990) offers a different explanation again, arguing from his own study of Winchester that low productivity was in fact deliberately tolerated in exchange for stability and sustainability. Looking at the broad picture, the ecological crisis must be understood in relation to the social conditions through which peasant environments were organized. It seems perfectly reasonable to assert the general theory that an unrelenting extraction of surplus from peasant farmers led to agricultural stagnation. Postan finds evidence of some reinvestment of capital into agriculture, but this appears to have been sporadic and extremely limited. As an example of appropriation, peasants had to foot the bill for the construction of no less than 800 castles between 1066 and 1189 – the kind of capital expenditure which left little for reinvestment into agricultural environments. Indeed, if feudalism was threatened by an agrarian crisis, it was because the ruling class failed to invest in the forces of production.

By the end of the thirteenth century, feudal ties to the land were already showing signs of decline. Colonization of new lands, together with disafforestation following the 1217 Forest Charter, had led to an increase in the number of unbonded tenants. Also, the growing number of landless labourers meant that bonded labour was less necessary than it had been. Vassals increasingly accepted cash payments, instead of labour, from their villeins and used this money to secure full-time, waged labourers. By 1300, no more than 60 per cent of peasants were bonded to vassals, and only one in six were labouring on demesnes (Dodgshon and Butlin, 1990).

The contradiction inherent in an agricultural economy that failed to invest in agriculture led to a vulnerability to systemic shocks. The first half of the fourteenth century brought two catastrophic shocks, the like of which would have rocked even the most stable of socioenvironmental regimes. A population of six million in 1300 was to be reduced to between two and three million over the next three generations. In May 1315 a deluge began, with heavy summer rains destroying much of the year's cereal crops. The following year was no better and the resultant famine was experienced throughout Western Europe. The Great Famine was followed in 1348–49 by the Black Death, an epidemic of bubonic plague which killed perhaps as much as one-third of an already diminished population. The feudal reaction to the plague, and the more general impoverishment of the peasantry, was not always compassionate, particularly as the resultant labour shortage provided fresh reason for bonded labour. The Hundred Years War with France led to increased military levies while the Poll Tax of the 1370s provided further hardship and unrest. The Peasants' Revolt of 1381, led by Wat Tyler and John Ball, was not primarily a call for social equality, but it was a cry for liberty and justice, and included the call for an end to bonded tenancies. While this revolt was crushed, it marked another step towards the end of feudalism and the emergence of democracy.

4.8 Nature in the medieval cosmology

The early medieval period is sometimes referred to as the 'Dark Ages', owing to a technological and intellectual stagnation. However, medieval Europe did experience significant currents of intellectual enterprise and these contributed to a changing understanding of nature. The dominant intellectual current was a theological one, embedded in the Church's struggle to defeat paganism and assert its right to challenge royal power. The key theological development was a coming together of scientific rationality and Christian faith, a synthesis that was to result in a period of relative harmony between church and science. The dispersion of new ideas was facilitated by the institutionalization of education, first in the monastic schools and later on in the universities. In England, Oxford University was founded in the late twelfth century, shortly followed by Cambridge. Less orthodox intellectual currents were also disseminated. The radical Christianity of St Francis of Assissi spread rapidly in the first half of the thirteenth century, while the mystic revelations of Hildegard of Bingen (1098– 1179) preached of a female side to the Christian god, sometimes interpreted as 'mother nature' (Ward, 1994). This chapter proceeds with a summary of the dominant beliefs about society and nature in the late medieval period and concludes with a consideration of the relationship between these beliefs and actual environmental behaviours. The question of this relationship was first broached in the conclusion to Chapter 3.

Four important trends within medieval cosmology are worthy of mention here. Firstly, the location of a single god in a heavenly sphere involved a separation of the spiritual domain from the earthly domain, a prelude to the more radical separation of society from nature described in the next chapter. Secondly, the Christian belief in God's creation of the world led to a **teleological** view of nature. This means that nature was viewed as having been created and designed in a way which served God's purpose. The workings of nature could therefore be explained through reference to God's masterplan and, by the same faith, understanding nature could provide insight into what God's masterplan was: the Bible could lead us closer to God, but so too could scientific enquiry. Thirdly, there was a continued link between economic interests and intellectual projects. The interplay between dominant ideas and dominant classes saw feudalism's justification sanctified in the prevailing religious ideology. Fourthly, the synthesis of science and theology included the acceptance of a scientific explanation for the superiority of men over women. Aristotle (384–322 BC) was undeniably a great thinker but, unfortunately, believed that women were incomplete and naturally less capable than men; even more unfortunate was the adoption of this belief by religious leaders.

Science and theology were harmonized through the synthesis of Aristotle's science with the Bible, as exemplified in the work of St Thomas Aquinas (1225–75). Aristotle's universe was composed of one basic material which had four qualities: hot, cold, moist and dry. These combined to produce four elements: earth, air, fire and water. Earth was at the centre of the universe, surrounded by a sphere of water, then a sphere of air and finally by a sphere of pure fire (Reed, 1990; Pepper, 1984). This was the world below the moon, which marked the beginning of a series of celestial spheres. Theological incorporation led to an ingenious, teleological, explanation of nature's processes. To the astronomy of the day, the celestial sphere was constant in motion and highly ordered: a clear indication of two of God's criteria for perfection. It therefore followed that the relative chaos of movement on earth was an aberration, explained by a temporary turmoil. In this explanation, an initial order of the elements had been disturbed by the introduction of motion; hence the observation that in some places earth was to be found above water, fire under the earth, water above fire, and so on. Nature was far from its intended state and God's design was for the elements to be restored to their original order. Within this framework, observations of nature could be easily explained: if you dropped a clod of earth, it fell downwards, demonstrating its purpose of returning to the centre of the universe; likewise, if you lit a fire, the flames would rise upwards, seeking their designated position above earth, water and air. Some water rose from springs, finding its way up through the earth to its proper position; other water fell through the air to do likewise.

Pepper (1984) highlights two more significant components of the medieval view of nature. Firstly, in line with the acceptance of Aristotle's natural science, the Bible was no longer the sole authority from which to seek knowledge. Nature itself became a 'book' which one could read and learn from. Secondly, there was an organic view of nature in which all components of nature (including humans) were connected in a Great Chain of Being, a view which certainly seems to place society and nature in a close and fragile relationship, something akin to the interconnectivity encountered in modern ecosystem ecology. The consequence of being part of the chain of nature might be a profound interest in keeping the links of the chain intact.

Over the last decade or so, academia has increasingly turned its back on arguments which begin with a description of apparently nature-friendly beliefs (such as the Great Chain of Being) and proceed to an assertion that a society 'lived in harmony with nature'. Such arguments have tended to romanticize the past, painting an unrealistic picture of society–nature relationships. Evidence consistently suggests that apparently respectful cosmologies can be compatible with significant and sometimes unsustainable environmental modifications. For example, it is widely (but not unanimously) thought that hunter-gatherers 'overkilled' certain large mammals, leading to many extinctions during the Pleistocene Era. With similarly animistic beliefs, neolithic farmers burned and cleared large tracts of forest, sometimes resulting in high levels of soil erosion. Even when we look to Eastern religions such as Taoism and Buddhism, the deepest respect for nature appears capable of supporting considerable modification for human benefit. Reviewing the environmental impact of medieval times provides yet another example of an apparent difference between belief and behaviour. Environmental behaviour does not seem to have been much constrained by a belief in a Great Chain of Being: the wolf, for example, was hunted as a pest without any legal intervention.

Hughes (1994, p. 35) offers one possible explanation for this divergence between cosmology and actions: 'It is significant that the first urban societies were also the first societies to abandon a religious attitude of oneness with nature and to adopt one of separation. The dominant myth and reality in Mesopotamia was the conquest of chaotic nature by divine–human order.' The crux of this argument comes in Hughes' implicit assertion that the way in which societies use nature *precedes* the way in which they think about it. Here, the forces and relations of production advance to the point where they are incompatible with the prevailing cosmology (of oneness with nature) and it is the cosmology which has to change (to separation). We might employ this thesis to explain the process of evolution in late medieval society: the feudal way of exploiting nature eventually led to the separation of nature and society which came in the early modern period: practice *preceded* belief. This linear, cause-and-effect explanation asserts that changing technologies and production methods are *the cause of* changing ideologies. David Pepper, for one, is roundly scornful of such **technological determinism**, an indication of the lively debate that continues to surround explanations of social change.

A second attempt to explain the difference between medieval belief and practice makes reference to the importance of **dominion** in Christian thinking:

> And God said, let us make man in our image, after our likeness: and let them have dominion over the fish of the sea, and over the fowl of the air, and over cattle, and over all the earth, and our every creeping thing that creepeth upon the earth . . . (Genesis, 1: 26).

White (cited in Attfield, 1991) considers dominion to be the historical root of our ecological crisis: a view which has been widely debated. A belief in dominion certainly subordinates nature to human needs but dominion can be interpreted as a caring form of stewardship as easily as it can be interpreted as a call for tyrannical exploitation.

What we can say with some certainty is that all premodern societies have had their ecological difficulties and that a deep respect for nature is not a simple solution to this. Harvey (1996, p. 188) suggests that there is no real incompatibility between a society's 'closeness to nature' and 'ecosystemic transformations that undermine their ability to continue with a given mode of production'. The ability to translate spiritual beliefs into perfect environmental management strategies is simply beyond the ability of past (and present) societies, particularly when a society is growing and having to intensify its use of nature. Nature is too complex and unpredictable to be perfectly managed and, sooner or later, societies are faced with the *unintended* consequences of their actions. (This is not a call for complacency!)

4.9 Further reading

For greater detail on early cities, see Morris (1994) *History of Urban Form*.

For the geography of feudalism in Britain, consult either Dodgshon and Butlin (1990) *An Historical Geography of England and Wales*, or Reed (1990) *The Landscape of Britain*. David Pepper (1984) provides a useful introduction to medieval cosmology.

A couple of books demonstrate modern examples of hydraulic despotism. If you were interested in Wittfogel's thesis, check out these more contemporary works: Donald Worster (1992), *Rivers of Empire*, investigates the control of water resources in the development of the American West, while the first essay in Arundhati Roy (1999), *The Cost of Living*, examines the controversial damming of the Narmada River.

4.9.1 Websites

For updates on the excavation of Hamoukar, visit the website of the University of Chicago's Oriental Institute: http://www-oi.uchicago.edu/

Chapter 5

Early modern environments

5.1 Introduction

While historians generally consider the early modern period to encompass the years from 1500 to 1700, we will extend it to 1800. It then includes the scientific, agricultural and English revolutions of the seventeenth century, the Enlightenment of the eighteenth century and culminates in the French Revolution of 1789 and the English industrial revolution that began in the 1780s. This period saw the establishment of a civilization and system of states in Western Europe that became associated with modernity and the West, and that were later to gain world dominance. By rejecting its own traditional culture, the West laid the foundations for a common civilization based on new social and environmental relations, and the new forms of development associated with modernization. These opened up a great technological and material gap between the West and the rest of the world, compelling most other states eventually to modernize themselves. The spread of modern institutions, ideas and techniques had profound effects on the material construction of nature as environments and on its conceptual construction within social and political thought. This chapter examines early modernity's impact on nature and environmental thought with particular reference to Britain and Europe.

5.2 The 'bourgeois revolution'

The rise of modern societies required a social revolution that established new forms of economic, political, social and intellectual organization and allowed the development of new technologies. This 'bourgeois revolution' occurred at different speeds, in different ways, in different countries, and involved three main elements (Gamble, 1981):

- **New ways of organizing economic production and distribution that allowed the rise of capitalism** These involved the gradual organization of a world economy and market, and the transformation of the relations between those who owned and worked the land. The voyages of exploration and conquest that began at the end of the fifteenth century opened the world to Western trade, boosted commodity exchange, fostered capital accumulation and the development of a financial system, and encouraged the application of new technologies. New social relations in the countryside released serfs from their feudal obligations and from the land itself. As large areas were enclosed and farmed more intensively by their owners, production shifted from self-sufficiency towards the market. These changes produced an agricultural surplus and a supply of 'free' labour available to work for wages in urban factories.

- **The development of the modern state** This originated in the long conflict between the Church and royal power, and eventually resulted in the organization of states whose authority was entirely secular. They were all powerful within a given territory but their power was no longer identical with that of any single individual. Public power became organized through the permanent agencies and institutions of the state and these facilitated the rise of free markets in land, labour and goods of all kinds. They did this by providing citizens with rights and responsibilities that allowed the growth of **civil society**: a sphere of free competition and exchange, in which individuals could pursue their private interests; forming associations and making contacts as they chose, free from detailed regulation and restraint by governments.

- **The rise of science and the modern ideology** Modern science became possible once a new cosmology separated nature from God and society and presented it as an alien realm of dead matter to be observed, analysed, appropriated and mastered by human intelligence (mechanical materialism). The development of empirical and rationalist modes of thought, and their application to technology and bureaucracy, radically changed the theoretical and practical relationship between societies and nature. Modern science provided the knowledge needed to run the new forms of economy and state and elements of the modern ideology needed to justify the new kinds of progress. Communities of intellectuals shaped this ideology which contained both a set of moral discourses and practices concerned with the good society, right order, social justice and the full development of human potential, and a set of rational discourses and practices that are concerned with the question of truth and knowledge, and with the discovery of the most effective means to realize given ends.

While the term 'bourgeoisie' was first used to refer to those who had given up active employment to live off the interest from their capital, revolutions in Europe rarely entailed an independent bourgeoisie simply overthrowing the aristocracy and its state. They were generally more complex affairs with elements of the new class having to gain support both from within the old ruling class and among the ordinary people. The English civil war in the seventeenth century shifted the balance of power between the king and the property-owning classes. A parliament dominated by landed interests then passed laws favouring the interests of liberal aristocrats who were dismantling the feudal system and encouraging new progressive tenant farmers. The rise of capitalist agriculture, or the agricultural revolution, together with increasing overseas trade, were the main sources of Britain's wealth in the eighteenth century and modern agriculture radically transformed the relations between people and the land (Shoard, 1987).

5.3 Capitalism

Like other modes of production, capitalism is a distinctive way of organizing the labour process or the constant interaction between people and the rest of nature (Johnston, 1989b; Lekachman and van Loon, 1981; Onimode, 1985). What makes it distinctive are its social relations whereby one class (**the bourgeoisie**) owns the means of production, distribution and exchange, while the other (**the proletariat**) has only its labour to sell. These class relations mean that by virtue of their monopoly, the former can exchange the goods and services produced with their resources at a value in excess of that representing the workers' wages (the labour value required for workers' subsistence). The main purpose of the bourgeoisie or capitalists is to accumulate this surplus value: the difference between the exchange value of commodities in the market and the labour value required to produce them being the primary source of their profit. Such accumulation is essential because capitalism is a competitive system. Capitalists who are unable to accumulate and reinvest sufficient capital will be driven out of business by their competitors.

The mechanical clock and mapping based on new surveying techniques, allowed modern societies to replace local time and space with **universal time and space**. This meant that economic acts of production, exchange and consumption could more easily be separated in space and time, and that goods and services could become commodities, acquiring a generalized exchange value (a price in the market) that is generally unrelated to their use value. Marxist economists employ a three- rather than four-capital model of the economy (p. 11), using **forces of production** (land, labour and capital) to refer to those elements of ecological, human and manufactured capitals that are used directly in the production process. For them **capital** only refers to such things as buildings, machines and stocks and shares that result from and represent the investment of money or surplus value created by employing labour. By referring to **conditions of production** they recognize that as well as the forces of production used directly in production, capitalism requires other elements of ecological, human and social and organizational capital if it is to reproduce itself in a sustainable manner. These include ecological resources and services not directly consumed in production (such as fertile soil, clean water and climate regulation); urban and rural space free from pollution and congestion; and human health and welfare (such as healthy and suitably educated young workers).

Marx's account of capitalism suggests how ecological and social relations are combined to produce commodities for sale in markets. This is summarized as **the circuit of capital** (Figure 5.1a) in which money purchases means of production; employs them in the production process to produce commodities; sells these commodities in markets for consumption; and then either recycles the resulting money back into the process (purchasing new means of production) or takes it as profit for the investor. A more sophisticated view of capital

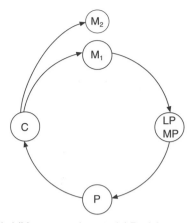

The circuit of capital (M_1 = money invested, LP = labour power, MP = means
of production (including raw materials from nature), P = labour process,
C = exchange of commodities, M_2 = profit)

Figure 5.1a The circuit of capital
Source: Dickens, 1996.

circulation (Figure 5.1b) suggests not one but three circuits of capital. In
addition to the primary circuit enabling commodity production, additional
circuits allow investment in fixed capital, in science and technology, and in
human labour. Investment in such fixed capital as office blocks, retail parks
and transport systems is made in expectation of profits via rents or accumu-
lating values, while investment in science, technology and education is made
in expectation of increases in productivity and hence profits. All circuits have
the effect of incorporating nature into society and creating by-products that
may impact on the workings of ecological and social systems. They may serve
to reproduce or undermine the conditions of production and so render capitalism more or less sustainable.

Capitalism's dynamism results from competition between capitalists and
related pressures constantly to increase the productivity of natural resources
and services used in the production process. The constant drive to increase the
productivity of labour and land leads to the exploitation of both human and
non-human nature, while failure to invest in the conditions of production
leads to such problems as water pollution, urban congestion and disease. By
undermining its forces and conditions of production, capitalism creates **two
contradictions** that shape its continuing development. The contradiction
between the forces and relations of production leads to workers seeking
greater control over the use of land, labour and capital, while that between the
conditions of production and the social relations that reproduce these conditions, leads citizens to demand greater control over the conservation of nature,
the planning of urban and rural space, and the promotion of human health
and welfare. Workers' and citizens' movements have won significant victories
over capitalists and their political allies in modern times, but such gains are
threatened during periods of overproduction or crisis.

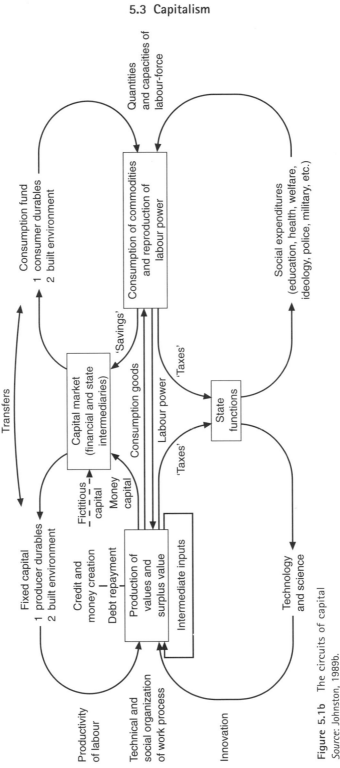

Figure 5.1b The circuits of capital
Source: Johnston, 1989b.

Because capitalism is a competitive system with no overall regulation or control, it contains the seeds of instability or **economic and social crisis**. Capitalists will continue to invest in greater output and drive down wages until markets are saturated and/or workers have no money to buy the goods they produce. Such a crisis of overproduction, overaccumulation or underconsumption may be compounded by an environmental crisis if capitalists also fail to reproduce the conditions of production in a sustainable manner. It means that profitability declines and there are contradictory pressures to cheapen raw materials and labour further, to increase workers' wages and consumption, to make production more efficient, to find new sources of raw materials and markets, and to invest in the conditions of production. If the response to these pressures is not successful then capital has to be scrapped and investment switched to new forms of production involving new technologies and labour processes. Such **restructuring** has profound implications for workers and environments as it reshapes social and environmental relations. People and environments in one location suffer the costs of factory closures, job losses and dereliction, while those in another are required to relate to one another and the rest of nature in new ways. Capitalism continually makes, destroys and remakes societies, places and environments, with major crises of profitability and subsequent restructuring occurring roughly every 50 years. We return to capitalism's uneven development through time and across space in Chapter 6.

5.4 Capitalism shapes the land

Capitalism replaced feudalism in Western Europe once there was surplus wealth to be invested in new ways, and changed social relations allowed this to happen successfully. It arose in the city-states of Renaissance Italy and spread northwards with the growth of trade and the increased use of money as a common means of exchange and reliable source of value. Trade and handicraft production created a new class of bourgeois entrepreneurs in the cities and the influence of the capitalist market gradually spread to the surrounding countryside.

As we saw in the last chapter, feudal villeins were granted rights to use village land in return for work and taxes owed to the manor. They were not allowed to sell produce or make loans in ways that allowed them to accumulate wealth and all economic exchanges were based on custom or face-to-face bargaining. Production was largely for subsistence and village life was largely self-sufficient. Norman kings provided a sophisticated armoury of local by-laws to protect commoners' rights to use a significant proportion of parish land and ensure its conservation. Common land provided the poor with some protection against strong landowners who considered their land as a trust to be exercised on behalf of the king, rather than a source of wealth and income.

The arrival of agrarian capitalism radically changed the land regime. Landowners under pressure to support wars and meet the costs of their own extravagance, gradually allowed an increasing number of serfs to become tenant farmers. Some tenants used new technologies to increase the productivity of the land greatly, with the result that rents rose and others were encouraged to follow their example. The dissolution of the monastries in the mid sixteenth century, and the confiscation of Royalist estates by Cromwell in the mid seventeenth century, added to the number of private landowners who regarded land as an investment and source of income. It was their interests and power that resulted in the enclosure of common land which reached a peak in the late seventeenth and eighteenth centuries (see Box 5.1). The

---

Box 5.1   Enclosure

The emergence of modern capitalism required peasants living under feudalism to be freed from the bonds and duties that tied them as forced labourers to aristocratic landowners and particular rural locations. It was the enclosure of open fields and commons, on which peasants previously had rights to grow crops, graze animals and collect wood and other products, that effectively turned them into free labourers. They were then forced to move in search of work and it was their cheap labour that allowed the growth of factory production and the industrial revolution. In the twelfth century villagers had rights over half the land of Britain. It was unfenced and worked and managed in largely collectivist ways. By 1876 less than 1 per cent of the population owned 98 per cent of the land and the vast majority had no land rights.

Enclosure dates from the fifteenth century when sheep-rearing and wool production began to be more profitable than arable farming. Early enclosures were concentrated in a belt running from Yorkshire to Gloucestershire and by 1720 around half the open field area had been privatized by private agreement, sometimes confirmed by court decree. Enclosure allowed private landowners and tenants to profit from new agricultural technologies such as crop rotation and stock-breeding, and was often accompanied by the clearing of woodland, the drainage of wetlands, or the creation of deer-parks for hunting. It was opposed by governments until the early years of the seventeenth century due to fears of rural unrest and depopulation, but Cromwell's parliament accelerated enclosure by abolishing feudal tenures and giving landowners greater powers over their estates. In the eighteenth century parliaments dominated by landowning interests passed Enclosure Acts which the poor had few powers to resist. These caused the enclosure of more than seven million acres of land, one-third of which were common pasture and waste. As the poor lost such rights as those to graze animals, cut turf, gather wood and collect berries, the subsistence economy that supported them was effectively destroyed (Figure 5.2). Many smallholders were left with parcels of land too small to provide them with a livelihood and many were forced to sell out to richer neighbours. While the rich increased the value of their estates considerably, poor relief were designed to make life so intolerable for rural paupers that they were forced to migrate to any job available.

There were hundreds of riots and revolts against enclosure in the sixteenth and seventeenth centuries that were brutally put down by landowners and their armies. The English Revolution benefited landowners and the new bourgeoisie, rather than the poor, encouraging dissenters to develop coherent

(continued)

(continued)

Figure 5.2   Enclosed land on Exmoor.
Source: Atmosphere picture library.

critiques of the prevailing rural order and blueprints for a very different England. Groups such as the Diggers and Levellers used the relaxation of censorship and publishing laws between 1640 and 1660 to challenge old institutions and beliefs, and the new class of free labourers was particularly receptive to their ideas about democracy and land reform. The Diggers sought equality and land rights for all, and began to dig up the commons in protests designed to draw attention to their underuse at a time of high food prices and rural poverty. The occupation and cultivation of St George's Hill, near Walton on Thames, in 1649 was harassed and eventually dispersed, but it served as an example for many similar occupations throughout the country.

Gerald Winstanley was the Levellers' leading thinker and writer. His book, *The Law of Freedom*, questioned the basis of private property, suggesting that while God had given the earth to everyone equally as a common treasury to benefit all, landowners or their ancestors had seized land by force. He believed that the problem of poverty could be solved by the common ownership and collective cultivation of land then lying idle, but sought such change through persuasion, universal male suffrage and education, rather than through bloody revolution. Winstanley influenced Thomas Paine and the American Revolution, but his ideas faded from public attention as the bourgeoisie made use of liberalism as a powerful way of accommodating radical dissent.

Sources: Shoard, 1987; Rackham, 1994.

agricultural revolution involved an increasingly industrial technology being used to work, drain and fertilize the land, with the emphasis on grain for subsistence shifting to livestock-rearing and root crops, as sheep-rearing and the export of wool became more profitable (Blunden and Turner, 1985; Hoskins, 1955).

The premodern or medieval economy was based on organic materials and renewable energy sources (wood, water, wind and draught animals), whereas the new capitalist economy was increasingly based on inorganic materials (metals) and non-renewable energy (coal) (Figure 1.2, p. 7). This changed pattern of resource use had a major impact on the environment, depleting forests of timber for charcoal, ship-building, and glass and soap manufacture in such areas as the Weald, and then creating new mining and industrial areas in South Wales, Shropshire and the Midlands. The industrial revolution after 1750 developed yet more new sources of fuel and supplies of metal, and harnessed new technologies such as the steam engine to the production of such commodities as cotton textiles and pottery for which there were huge increases in demand.

5.5 The modern state

As a mode of production based on class relations, capitalism requires the protection of privilege or the safeguarding of private property via the political, legal and ideological organization of society. This function is performed by the modern nation-state that evolved alongside capitalism and civil society as the power of absolutist states was challenged. Struggles between the Church, the monarchy and the people, gradually resulted in the state emerging as an agent of the new bourgeoisie, enabling the latter to throw off absolutist and feudal fetters and to promote its own interests as the general interest. Inspired by the idea of free and equal citizens who could manage their personal lives as they wished and decide on matters democratically, the bourgeoisie called for 'liberty, equality, and fraternity'. At the same time it promoted the market economy as a sphere of freedom and equal opportunity in which all citizens could realize their own objectives without interference of the monarchy, Church, estate or guild. Capitalism was thus the first mode of production which its advocates and defenders proclaimed to be classless and capable of promoting the interests of all on the basis of equality.

The modern state acts as a centralized power with supreme authority to make and enforce laws and to regulate social arrangements and relations within a given territory. It is both a product and a guarantor of civil society, guaranteeing citizens civil, political and social rights in return for responsibilities, and exercising its power through such institutions as the bureaucracy, judiciary and military. Governments merely run the state for a fixed term, and in a liberal democracy the legitimacy of the state and the popularity of governments depend on their ability to appeal to different interests and satisfy a sufficient number of the electorate. The modern state is founded on consent between those who control its institutions and those who are subject to its regulation, and this relationship is more or less stable at different times and in different places. The modern period has seen an increase in the powers and provisions of the state; a widening of its social basis or the franchise; an extension of the rights and responsibilities of citizens; and greater use of the state as an agent of modernization or development. Its role in regulating environmental relations has also increased significantly.

To understand how the modern state shapes the social use of nature and the environment, it is necessary to consider its three core functions (Johnston, 1989b). Firstly, the state regulates and sustains the market system. It does this by protecting private property rights and ensuring contract compliance; providing a common currency and weights and measures; and ensuring the reproduction of such conditions of production as clean water, transport and basic education. It also manages periodic economic crises by, for example, helping capital to find cheaper sources of raw material, subsidizing research and job creation involving new technology, or becoming a major purchaser of infrastructure projects or armaments.

Secondly, the state regulates social or class conflict. It counters the potential mobilization of workers and citizens in support of programmes designed to bring about capitalism's downfall, by meeting some of their demands for civil, political and social rights. As rights and associated responsibilities expand, the state may come to be seen as the champion of the less privileged by providing, for example, rights to protection in the workplace, political representation and clean food and water. Demands for the protection of public health via water-treatment systems was a major factor in the creation of local states or governments in Britain, and while such expenditure increases the tax burden on some capitalist enterprises, it provides orders and profits for others.

Thirdly, the state secures social order and consensus by requiring citizens to accept the rule of law and be subject to the kind of ideology that the state promotes. Laws govern the ownership and use of land and natural resources, and the rights of owners have gradually been constrained to reflect wider interests. State-provided or sponsored education systems are one means of promoting ideology supportive of existing social and environmental relations, with much that happens in schools, colleges and universities encouraging students to accept capitalism and the capitalist state as normal, inevitable and in everyone's interests. Education as praxis (p. 39) is relatively rare. Few students are encouraged to develop a critical understanding of society and nature in the classroom, and the ideology they encounter there is reinforced by much that is communicated or 'taught' through the mass media.

5.6 The early modern state in England

In the early modern period the groups that made and administered the law in England (the executive and judiciary) were dominated by landowners and their interests. The ownership of property was the key to joining the electoral register and standing as a candidate, and only a tiny fraction of the population could vote (agricultural workers did not gain the right to vote until 1884; town workers gained it in 1867). The distribution of constituencies favoured rural areas, the ballot was not secret, voters were often bribed or coerced by landowners, and politics had more to do with the relations among landowners than class conflict. Liberalism and natural law (see below) were used to justify the right to private property and its protection under the law, and the Game Law of 1671 provides just one example of how nature was appropriated by the rich. It forbade all without property to take game, outlawed the possession of dogs, traps, etc., and so cut off a major source of the poor's food. It was harshly applied and strongly opposed, and led to the Black Act of 1723 that introduced the death penalty for taking deer, rabbits and fish. Similar laws restricted access to the countryside, confining those without land to predetermined routes and preventing their enjoyment of many traditional forms of recreation.

5.7 Modern science and rationalism

The philosophers whose ideas shaped the scientific revolution of the seventeenth century and the Enlightenment of the eighteenth century (see Chapter 2) realized the connections between their new mechanistic view of the world and the needs of the new bourgeois entrepreneurs. Modern science, based on empiricism and rationalism, developed alongside such technologies as navigation, gunnery and hydraulics, and helped to systematize and universalize techniques and values appropriate to the new capitalist order. Science as a system of rational knowledge, constantly subject to review and change, reflected and responded to capitalism's rationalization of social and environmental relations and its constant need to revolutionize the means of production. Science became integrated with the production process and its mechanistic framework legitimated the transformation and domination of what was now perceived as a dead and disenchanted nature.

The impact of the scientific revolution and Enlightenment spread beyond the economy to foster an age of reason in which all forms of authority were increasingly subject to sceptical questioning, nothing was taken for granted, and rational criticism, rather than censorship, was considered the cure for wrong ideas. **Rationalism** became the foundation or guiding principle for all knowledge and an emerging modern faith in reason and science meant that people increasingly believed in a rational world that could be analysed, understood and changed through human action. The natural and social worlds were rational to the extent that people had the knowledge and will to control them in accordance with their own purposes. Obstacles to such control were considered irrational and to be swept aside.

Rationalism became associated with a set of modern moral values and a set of technical means that were to prove increasingly contradictory as capitalism and the modern state developed. Modern values suggested that individuals should be free to act as rational independent beings; that all social arrangements should be justified by the standards of reason and should contribute to the liberation and all-round development of each individual's capacities;

and that the natural world should be mastered and subordinated to human purposes. Modern technical means required action to be governed by a search for the most effective means of reaching a specified goal with the search conducted rationally, according to scientific method. Under the social and environmental relations associated with industrial capitalism and the increasingly bureaucratic modern state, technical or instrumental rationality elevated efficiency and means over values and ends, with the consequence that modern ideals of individual autonomy, human emancipation and mastery of nature were not fully realized. The Western ideology had to accommodate this and other emerging contradictions, with liberalism and socialism providing the basic vocabularies and perspectives for responding to the social and environmental impacts of modernization.

5.8 Liberalism and egocentric environmental ethics _____

The early modern period saw the rise of **liberalism**, an ideology associated with the rise of the new bourgeoisie and their desire for freedom from remaining feudal and monarchical control. From supporting basic civil liberties or human rights (liberty, equality, fraternity), liberalism developed into a political creed stressing the independence of ordinary men against any powerful body such as the state or organized labour (Heater, 1974; Jones *et al.*, 1991). Liberalism attributes its own meanings to such modern ideals as freedom, equality, rationality and sovereignty and, as suggested above, enables the advocates and defenders of capitalism to promote their own interests as the general interest. It is strongly associated with *laissez-faire* capitalism or absolute opposition to any government or state control of aspects of the economy. Optimum economic performance and maximum social good results from everyone being subject to the power of market forces. Attempts to regulate free markets, for instance by banning child labour, allowing collective wage-bargaining or charging for waste disposal, should be resisted in favour of total economic freedom.

Liberalism and the liberal social theories that developed in the early modern period, reflect mechanistic science in the following ways:

1. Mechanistic science is based on the assumption that matter is made up of individual parts. Atoms are the real components of nature, just as individual humans are the real components of society.
2. The whole is equal to the sum of the individual parts. Similarly, society is the sum of individual rational agents.
3. Mechanism involves the assumption of context independence. Real objects obey the laws of falling bodies and gravitation only when environmental contexts such as air resistance and friction are stripped away and masses act as point centres of force. In society, rules and laws are obeyed by a populace comprising equal individuals, stripped of particularity and difference.
4. Change occurs by the rearrangement of parts. In the billiard ball universe of mechanistic scientists, the initial amount of motion (or energy) introduced into the universe by God at its creation is conserved and simply redistributed among

the parts as they come together or separate to form the bodies of the phenomenal world. Similarly, individuals in society associate and dissociate in corporate bodies or business ventures.

5. Mechanistic science is often dualistic. [*Early modern scientists*] posited a world of spirit separate from that of matter. Nature, the human body, and animals could all be described, repaired and controlled, as could the parts of a machine, by a separate human mind acting according to rational laws. Similarly . . . democratic society is a balance of powers as in a pendulum clock, and government operates as do the well-oiled wheels and gears of a machine controlled by human reason. Mind is separate from and superior to body; human society and culture are separate from and superior to nonhuman nature (Merchant, 1992, pp. 68–9).

The model of the social world derived from mechanistic science was then one of self-contained individuals with built-in passions and drives who sought to maximize their satisfactions and minimize their dissatisfactions. Each was considered to have an equal right to life, liberty and the pursuit of happiness, and in striving to maximize his or her own satisfactions should respect the rights of others. **Utilitarianism** taught that in pursuing their own interests, individuals would enter into beneficial relations with others and that the totality of such arrangements would guarantee the greatest happiness of the greatest number of people. Such ideas underpin technocentric environmentalism (p. 19) and an **egocentric environmental ethic** based on self-interest that allows individual entrepreneurs and corporations to extract and use natural resources for profit while claiming that such use is also in the wider public interest. This ethic also draws support from interpretations of Judeo-Christian teaching that claim people have a mandate to use nature for their own purposes ('Be fruitful and multiply and replenish the earth and subdue it', Genesis, 1: 28) and from the Protestant ethic with its twin imperatives to methodical work as the chief duty of life, and to the limited enjoyment of its product. The unintended consequence of reformed Christianity's 'worldly aestheticism', enforced by social and psychological pressures on the believer to prove his salvation, was the accumulation of wealth for investment.

5.9 The role of intellectuals

Early modern intellectuals were remarkable for the breadth of their scholarship or their ability to develop new ideas across the natural and social sciences. It was still possible for one person to grasp most of what was known about the world, and while there were the beginnings of specialist subjects such as geography (see Box 5.2), academic divisions of labour hardly existed. Many early scientists were also engineers and philosophers and they contributed to a series of national enlightenments, in France, England, Scotland and elsewhere, that constituted a sharp break with premodern or traditional ideas. Such enlightenment did not, however, entirely discard the classical and medieval legacy. The Greek concept of the city-state, citizenship and a public realm of freedom beyond the necessities of everyday life continued to influence modern thinkers, as did the Christian concept of individuals, each with an immortal soul, all equal before God. It was the role of key intellectuals to revisit and revise such ideas, incorporating them, along with new ideas, in the diverse doctrines, such as liberalism and socialism, that comprised the Western ideology. These doctrines were concerned with the new economy, the new state and the new science, and were much influenced by differing ideas about nature.

In his *Wealth of Nations* (1776) the political economist Adam Smith suggested that individuals pursuing their own interests in conditions of free competition produced not only a 'natural' social order but also the most rapid possible increase in people's well-being and the wealth of society. He contrasted the natural order emerging with capitalism with the artificial orders imposed by tradition, aristocratic interests or ignorant meddling, and suggested that it could be scientifically proven that the capitalist division of labour benefited all classes. Inequalities under capitalism were not incompatible with the natural equality of all men, or natural justice, for wealth 'trickled down' to the poor more effectively, and the free play of natural forces (equal exchange in the market) eventually destroyed all positions not built on contributions to the common good.

Political philosophers, who were no longer able to justify political arrangements by reference to tradition or theology, argued about **the state of nature**: an imaginary reconstruction of how life and social relations might have been before the creation of organized political society. While such arguments are entirely hypothetical, since politics has always existed in some form, imagined states of nature suggest the drawbacks of living in a pre-political environment and suggest what political arrangements would recommend themselves to people living there. In his *Leviathan*, published in 1651, Thomas Hobbes, suggested that people in the state of nature were unfriendly, violent and competitive. They had equal rights to the common property (or commons) that

---

Box 5.2 Two early modern geographers

Bernhard Varenius (1622–50) published his *Geographia Generalis* in Amsterdam in 1650. It contains much material that today would be labelled mathematical geography or astronomy. As well as sections that describe the shape and size of the Earth and its physical geography, there are sections that examine the relations between the Earth and other heavenly bodies, particularly the Sun, and discuss the location of different places in relation to each other and the principles of navigation. Varenius died at the early age of twenty-eight and did not have time to write his book of special geography. This would have described places in terms of their climate, fauna and flora, economy and form of government. By dividing geography into general and special branches Varenius anticipated what were later called systematic and regional geography. He sought to restrict general geography to physical conditions that could be understood through natural laws. He considered it difficult to establish laws in special (regional) geography and here description had to suffice.

(continued)

(continued)

Immanuel Kant (1724–1804) is best known as the philosopher who combined rationalism and empiricism to provide new foundations for modern knowledge (see Chapter 2). He believed that there are two ways of grouping or classifying empirical phenomena for the purpose of studying them: either by reference to their nature, or by reference to their position in time and space (Figure 5.3). The first method, logical classification, provides the foundations for the systematic sciences (e.g. geology, zoology, sociology), while the second, physical classification, provides the scientific basis for history (the chronological science) and geography (the chorological science). While Kant's approach to knowledge gave geography a central place within the sciences, few would now claim that it is possible to understand geography without reference to time or the systematic sciences. Kant's lectures on physical geography were very popular covering the races of human beings, their physical activities on the earth, and natural conditions in all their variety.

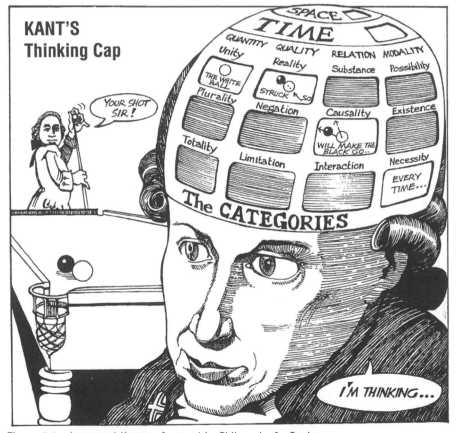

Figure 5.3   Immanuel Kant as featured in *Philosophy for Beginners*
Source: Osbourne and Edney, 1992.

Source: Holt-Jensen, 1988.

nature had given to all, but only a strong state with absolute power could guarantee such rights by enforcing a social contract that required rational individuals to overcome their 'natural' instincts to fight over property. John Locke (see p. 28) took a more liberal view of the state of nature, seeing it as one in which individuals are free, equal, interdependent, rational and moral beings. He believed that the unfettered acquisition of money, goods and land was part of nature and therefore ordained by God. Such acquisition was the main reason that people left the state of nature and set up government, but it should be free of all political and moral constraint as it was sanctioned by natural law. The state should therefore ensure citizens inalienable rights to life, liberty and, above all, property, with government subject to control by citizens who are able to remove a government should it exceed its limited powers. The ideas of Hobbes and Locke supported *laissez-faire* capitalism and an egocentric environmental ethic.

5.10 Changing attitudes to the natural world

The profound changes of the early modern period had a significant impact on ordinary people's attitudes towards the natural world. In documenting these attitudes, Keith Thomas (1984) points to a growing contradiction between increasing material wealth and comfort, derived from capitalism's exploitation of nature, and increasing concern for nature and the welfare of other species. He explains how popular assumptions about man's ascendancy over nature, which were almost unquestioned in Tudor and Stuart England, gradually gave way to doubts and hesitations as urbanization and industrialization intensified people's alienation from nature. By 1800 the growth of towns had led to new longings for the countryside, and the progress of cultivation had fostered a taste for weeds, mountains and untamed nature. Security from wild animals had generated an increased concern for wildlife conservation and nature parks, and conservation areas were beginning to become fantasies that enshrined the values by which society as a whole could not afford to live.

The growth of rural longings and associated pastimes illustrates these trends in popular thought. The early modern period saw a reversal of attitudes towards the town. With the growth of urban pollution and crime, and the disappearance of much green space from within urban boundaries, the town lost its associations with civilization, security and heaven. It became a place to be avoided, while the countryside was reconstructed as a virtuous, healthy and idyllic retreat. Artists and poets helped to engineer this reconstruction, but the new meanings attached to the countryside were essential evasions that ignored the harsh realities of rural life. They worked to ameliorate alienation and by the late eighteenth century nostalgia for rural England was reflected in such new pastimes as gardening, walking, wild-flower collecting, birdwatching and reading rural writers. By then many felt that although the natural world should be tamed it should not be completely dominated and suppressed. Such attitudes were to grow and contribute to the emergence of modern environmentalism.

5.11 Further reading _____

Chapter 6 'Social Ecology' in Merchant (1992) provides a sound introduction to socialist ecology and ecological Marxist theory. Try to relate Merchant's Figure 6.2, p. 148, to contradictions and crises you perceive in the world around you. Chapter 5 of Quaini (1982) also deals with capitalism and ecological contradictions.

In *The Song of the Earth*, Bate (2000) describes John Clare (1793–1864) as 'the most authentically working class of all major English poets'. Craig, in her review of Bate's book adds that 'for Clare, unlike middle-class Romantics, the bond with nature is the antithesis of escapism and the retreat from social commitment . . . his laments against enclosure and exile foreshadow scientific ecology in his knowledge that an organism has meaning and value only in its proper home, in symbiotic association with all the creatures that surround and nourish it' (Craig, 2000). Evaluate these claims by reading an account of Clare's life (e.g. Drabble, 1984, pp. 61–5; Grimes, 1995, pp. 271–90), some of his poems, including 'Enclosure', and relevant sections from Bate's book.

In Chapter 4 of *Imagined Country*, Short (1992) examines how the environmental myths of countryside and city are articulated in the environmental ideologies of the UK. Establish what Short means by myths, ideologies and texts (see his introduction), and then read the chapter paying particular attention to enclosure, landscaped gardens, and the city as both Babylon and Jerusalem.

Chapter 6

Ecological imperialism and the rise of geography

6.1 Introduction

This chapter explores how modernity and the West gained dominance over societies elsewhere in the world. It deals with the expansion of European influence from 1500 and the role of plunder, slavery and trade in enabling the industrial revolution. The nature and causes of imperialism, particularly ecological imperialism, are linked to capitalism's need to overcome economic and political limits to growth, and world systems theory is used to explain the emerging global division of labour. The rise of socialism as an anti-capitalist and anti-imperialist politics seeking to alter the direction of the modern project is then outlined. While some social scientists and geographers began to develop an understanding of society and nature based on Marx's dialectical materialism, mainstream geography remained dominated by simpler formulations. The chapter covers the period from 1500 to 1945 with a focus on the nineteenth century.

6.2 European expansion

European expansion took place in three phases. Between 1500 and 1700 the Spanish and Portuguese conquered Central and South America, while the British and French settled North America and extended trade along the African coast and into the Indian Ocean and Southeast Asia. This phase was essentially one of plunder, the accumulation of treasure, and the establishment of trade, particularly the triangular trade involving the transport of iron bars and manufactures to Africa, slaves to the West Indies and North America, and sugar and tobacco to Europe. In the second phase, from 1750 to 1850, the British defeated the French for the control of the Indian subcontinent; there was increased European trade with China; and settlement of Australia and New Zealand occurred. Colonies now provided cheap sources of raw materials and profitable markets for European exports. In the third phase, after 1850, Africa was divided between European powers; Russia expanded eastwards; and after the defeat of the Otterman Empire in the First World War, France and Britain established control over the Near East. Expansion and empire had by then become the means of exporting surplus population and investment capital. By 1914 European influence extended to 84 per cent of the world's land area; 500 million people lived under imperial rule, and Britain, France and Germany had invested $30,000 million in foreign and colonial loans (Crow and Thomas, 1994; Brazier, 1989; Hobsbawn, 1997; Ponting, 1991).

European expansion is best understood in terms of **imperialism**, the process by which nation-states extend their power or rule beyond their original boundaries by taking in other nations or lands and so creating empires. Empires may be formal if the conquered land is entirely absorbed into the administrative and political institutions of the imperial state, or informal if the constituent states retain some degree of independence and identity. **Colonialism** is a particularly exploitative form of imperialism in which only citizens or subjects of the 'homeland' are recognized as citizens of the empire, and the rest of the inhabitants are no more than subject people with no hope of political power or legal protection. A colony has no political independence from the 'mother country' and exists to increase the wealth and power of the imperial power by supplying cheap labour and raw materials, and acting as an avenue for profitable investment. The earliest colonies in the modern world, the British colonies in North America, India or Australia, for example, were set up by trading companies operating under royal warrants, with the express intention of making a profit. The desire to extend or defend empires has been a major source of conflict between imperial powers and one of the aims that Germany expressed in both world wars was the achievement of territory on a par with that of Britain.

Marxist political economists associate imperialism with the accumulation of wealth that enabled the rise of industrial capitalism in Europe and the process of expansion whereby capitalism is able to overcome economic and political limits to growth (Barratt Brown, 1974). Wealth accumulated through plunder, merchant trade and slavery, in the Americas, India and elsewhere from the sixteenth century, was concentrated in few hands and encouraged capitalist development by providing purchasing power, raising prices and profits in relation to wages and providing the funds for extended credit. While much of it was invested in land and grand houses, some helped to finance the agricultural revolution and so reduced labour costs. Yet more was made available to a new class of industrial capitalists to invest in new technologies, purchase labour and raw materials and build factories. When profits subsequently began to fall due to over-accumulation of capital, imperialism provided new sources of profitable investment in parts of the world where the rate of exploitation of human and non-human nature was higher than in Europe. At the same time cheap food and raw materials from the colonies contributed to the rising living standards of Europe's people and so served to reduce political unrest. At the start of the seventeenth century the wealth per head in India, China, Latin America and Africa was higher than in Europe, and these regions contained civilizations that were arguably more advanced than those to be found in Europe. In the subsequent 300 years their wealth was largely expropriated and their civilizations destroyed by processes that created what later became known as the 'Third World'.

6.3 Ecological imperialism

A key motive behind imperialism is the acquisition of new land and natural resources to incorporate as commodities within circuits of capital. Rates of return to capital invested in land on the 'frontier' outside the capitalist mode of production are often high and **ecological imperialism** refers to the process whereby imperial powers gain control of natural resources and services within their formal and informal empires. It occurs in two ways (Crosby, 1986). Either people from capitalist countries occupy the land themselves and impose their own form of social organization, as happened in the Americas, Australia and New Zealand, and parts of southern Africa, or lands formerly occupied by non-capitalist modes of production are slowly penetrated by the capitalist mode and so incorporated into the world economy. Both involve the destruction or degradation of local peoples and their culture, and forms of **environmental racism** that subject black people to the worst effects of environmental degradation and pollution. They destroy elements of the local fauna, flora and habitats; generally introduce less sustainable ways of life that intensify the contradictions of capitalist development; and prompt political and social movements urging independence and more sustainable forms of development.

Ecological imperialism within the formal empire of British colonialism is well illustrated by the establishment of pastoralism in Australia (see Box 6.1).

Box 6.1   Pastoralism in Australia

Australia is an especially old and fragile land, rich in biodiversity. The first humans arrived at least 50,000 years ago, and by about 20,000 years ago, people had spread to all parts of the continent. Aboriginal peoples were mainly hunter-gatherers with an intimate knowledge of the land who modified their environment, largely through the use of fire. With the arrival of Europeans, the rate of human-induced environmental change accelerated dramatically. Early settlers regarded both the environment and aboriginal people as hostile and alien. Aborigines were killed by introduced diseases, poisoned, hunted down, forced to retreat into the bush, and later confined on mission stations from where their children were stolen and 'adopted' into white society. At the same time the environment was gradually 'tamed' with introduced technology and alien species of plants and animals. The prevailing view of the environment changed from early dismay to unrealistic expectation, and it was not until the late twentieth century that most people realized its limitations and that its development had come at great cost.

Captain Phillips and the first fleet arrived in Sydney Cove in 1788, and after initial difficulties the colony of New South Wales grew to 10,450 by 1808. Despite the very obvious presence of aborigines, the British regarded the continent as *terra nullius*, an empty uninhabited land that could be progressively taken without negotiation. John Macarthur imported Spanish merino sheep in 1797, took samples of Australian wool to Yorkshire in 1803,

(continued)

(continued)

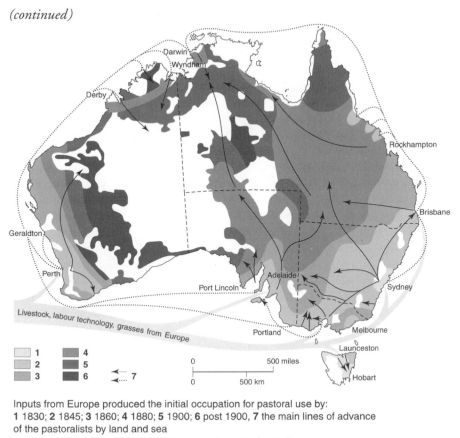

Inputs from Europe produced the initial occupation for pastoral use by:
1 1830; **2** 1845; **3** 1860; **4** 1880; **5** 1900; **6** post 1900, **7** the main lines of advance
of the pastoralists by land and sea

Figure 6.1   The expansion of pastoral land use in Australia
Source: Heathcote, 1975.

and set up the Australian Agricultural Company in 1824. The French Revolution and the British industrial revolution, together with the Napoleonic
wars, had produced a wool shortage in Europe and exports from Australia
rose rapidly. The first pastoralists were generally sons of aristocratic families
or yeoman farmers in England, who were granted land and convict labour in
return for capital investment in stock and assisted passages for new settlers.
By the 1820s resentment of their wealth and low stocking rates, led to the
abolition of land grants, an attempt to charge for grazing land, and the
further invasion of aboriginal lands by squatters and their stock. The squatter
era of 1820–40 saw the occupation of huge areas by illegal grazers, many of
whom were financially backed by the leading citizens of the colony (Figure
6.1). Government was powerless to turn them off, and by 1836 it was issuing
squatting licences for £10 a year. One £10 licence allowed James Wilerowang
to hold 27 stations and two million hectares of land.

(continued)

(continued)

Figure 6.2   Degraded pastoral land in Australia
Source: Still Pictures.

The lives of early squatters and their labourers were hard. Drought and falling wool prices brought many to poverty and ruin, and in the 1840s millions of sheep had to be boiled down for tallow. The 'squattocracy' was a powerful political lobby, arguing that it was only in the hands of private owners that the land would be properly developed and made most useful to the community at large. Squatters wanted security of tenure so that they could improve their runs and build homes without fear of being displaced. The Governor was reluctant to sell crown lands that would appreciate in value, but eventually legislation was passed in 1846/7 that gave the squatters what they wanted. They could get leases with pre-emptive rights to buy all the land they occupied and in the meantime could purchase key sites such as water-holes, river-beds and homesteads. Following the gold rushes of the 1850s that greatly increased the numbers and influence of the landless, there was pressure to open crown lands to freehold settlement. Land acts sought to break up large grazing leases into smaller selections, but the result was to concentrate ownership further.

Security of tenure promoted technological innovation and improvement. Fencing established boundaries and allowed greater control of stock: oil technology was used to drill for water in artesian basins; refrigeration reduced transport problems; and trawler tractors were used to clear huge areas for pasture. Many exotic plant species were introduced in attempts to improve

(continued)

(continued)

the quality of the pasture and the original stock breeds were greatly modified. The state played an increasing role in such improvement: building transport links; promoting migration and settlement; and providing extension services and price support.

Pastoralism has had a profound effect on the land. Hard hooves compacted loose soils while heavy grazing reduced grass cover and led to weed invasions. Extensive grazing by introduced species now covers 54 per cent of Australia and around 15 per cent of rangelands are sufficiently degraded to require destocking if the land is to recover (Figure 6.2).

Sources: Bolton, 1981; Powell, 1991; Shaw,1960; Wilson, 1987.

Since 1788, when the continent was invaded and claimed for Britain, almost 70 per cent of Australia's native vegetation has been removed or significantly modified, and aboriginal people have been subjected to systematic genocide. Decisions in Australia's High Court in the 1990s found that native title could coexist on pastoral leases, but subsequent legislation extinguished such title outright on all land other than crown land, and on a great deal of such land under pastoral lease.

6.4 World systems theory

Examination of pastoralism in Australia reminds us that there are no universal causes of environmental problems. Soil degradation and the loss of biodiversity over much of Australia resulted from processes of modern development working themselves out, alongside biophysical processes, in particular ways in particular places, at particular times. Chapter 2 suggested how critical realism offers a philosophical framework for understanding such problems by recognizing how biophysical and social structures and processes in the real domain (e.g. soil compaction, the global wool trade) are connected to experiences in the empirical domain (e.g. weed invasion of pasture, availability of land) through contingent factors or events in the actual domain (e.g. overstocking, passing of new land acts). The expansion of Europe intensified the process of **globalization**, whereby processes, events and experiences in one location are more and more affected by those in other locations elsewhere in the world. Societies around the world were increasingly integrated into a single mode of production and their environmental problems can only be understood by reference to the kind of understanding of the world economy offered by world systems theory.

A world system is a social unit larger than the local day-to-day activities of its members. Immanuel Wallerstein, the originator of **world systems theory**, suggests that historical societies can be classified according to their mode of production, into mini-systems, world empires and the world economy (Taylor, 1989; Wallerstein, 1984). Mini-systems are premodern, local in their range, and are dealt with in Chapter 3. World empires, the first type of world system, are also premodern, supporting a military-bureaucratic class from a surplus of production beyond their immediate needs, and producing the kind of environmental relations outlined in Chapter 4. The expansion of the European **world economy**, based on the capitalist mode of production, eliminated all remaining mini-systems and world empires and now links all the world's peoples and environments in a single system of combined and unequal development.

The world economy has three basic elements: a single world market, a multiple state system, and a three-tier structure. In the **single-world market** production is for exchange rather than use, and competition for market share largely determines the type, quantity, location and environmental impact of production. While there is one market there are many nation-states. If one state controlled the world market competition would be eliminated and the world economy would transform into a world empire. Powerful states, like Britain, the United States and Japan, within the **multiple-state system**, are able to distort the market in the interests of their national capitalists, and such distortion shapes the agenda of international politics. As states compete for economic, political and military power, the environment is inevitably affected and it too appears on international political agendas. Like any social system, a world economy is more stable, or less confrontational, if it contains three, rather than two tiers. Modern development produced a social and spatial division of labour, with three classes (working, middle, ruling) and a **three-tiered spatial structure** consisting of the core and periphery, with the semi-periphery separating these extremes of well-being.

World systems theory enables us to conceptualize the social use and production of nature taking place within a space–time matrix in which time is socially produced by the dynamics of the world economy and space is socially produced by its structure. The matrix (Figure 6.3) shows five waves of economic growth and stagnation in the history of modernity or the world economy, with the first long logistic wave, from 1450 to 1750, covering the decline of feudalism and the rise of the capitalist world economy resulting from the developments outlined above. The subsequent four Kondratieff waves (named after the Russian economist who first recognized them) are each associated with particular technological innovations and product cycles. Each creates distinctive environments, environmental problems and opportunities.

As the European world economy expanded it incorporated other societies, not as equals but on unfavourable terms. Imperialism established or reorientated economies and environments to meet European needs, subjecting them to processes that ensured their underdevelopment or peripheralization. Core and periphery refer to processes that produce development or underdevelopment, and not directly to areas, regions or states. Spaces become developed and core-like because of a predominance of core processes made possible by social relations that incorporate relatively high wages, advanced technology and a diversified product mix. Other spaces become underdeveloped and peripheral

	Core	Semi-periphery	Periphery
Logistic curve	A Initial geographical expansion based on Iberia, but economic advances based on north-west Europe. B Consolidation of northwest European dominance, first Dutch and then French–English rivalry.	Relative decline of cities of central and Mediterranean Europe. Declining areas now include Iberia and joined by rising groups in Sweden, Prussia and northeast United States.	Iberian empires in 'New World'. Second 'feudalism' in eastern Europe. Retrenchment in Latin America and east Europe. Rise of Caribbean sugar. French defeat in India and Canada.
Kondratieff wave I	A Industrial revolution in Britain, 'national' revolution in France. Defeat of France. B Consolidation of British economic leadership. Origins of socialism in Britain and France.	Relative decline of whole semi-periphery. Establishment of United States. Beginning of selective rise in North America and central Europe.	Decolonization and expansion – formal control in India but informal controls in Latin America. Expansion of British influence in Latin America. Initial opening of East Asia.
Kondratieff wave II	A Britain as the 'workshop of the world' in an era of free trade. B Decline of Britain relative to United States and Germany. Emergence of the Socialist Second International.	Reorganization of semi-periphery: civil war in United States, unification of Germany and Italy, entry of Russia. Decline of Russia and Mediterranean Europe.	The classic era of 'informal imperialism' with growth of Latin America. Expansion – scramble for Africa. The classical age of imperialism.
Kondratieff wave III	A Consolidation of German and US economic leadership. Arms race. B Defeat of Germany, British Empire saved. US economic leadership confirmed.	Entry of Japan and Dominion states. Socialist victory in Russia – establishment of USSR. Entry of Argentina.	Consolidation of new colonies (Africa) plus growth in trade elsewhere (especially China). Neglect of periphery. Beginning of peripheral revolts. Import-substitution in Latin America.
Kondratieff wave IV	A United States as the greatest power in the world both militarily and economically. New era of free trade. B Decline of United States relative to Europe and Japan. Nuclear arms race.	Rise of Eastern Europe and 'Cold War'. Entry of OPEC. Entry of 'Little Japans' in East Asia and new regional powers – China, Brazil, Mexico, India. Rise of debts to core.	Socialist victory in China. Decolonization leading this time to 'neo-colonialism'. Severe economic crisis and conflict. Expansion of poverty.

Figure 6.3 A space–time information matrix
Source: Taylor, 1989.

when shaped by peripheral processes and relations that incorporate low wages, rudimentary technology and a simple production mix. These processes operate in a single world economy and ensure that the core exploits the periphery and the periphery is exploited by the core. The nature of these processes is not determined by the product being made. In the nineteenth century peripheral processes were imposed on India so that it would provide Lancashire with cotton while core processes were gradually transplanted into Australia (see Box 6.1) so that it would provide Yorkshire with wool.

While India remained dominated by peripheral processes, Australia became a semi-peripheral space combining both peripheral and core processes. Its role within the world economy involved both exploiting the periphery (including its own internal periphery) and being exploited by the core. The semi-periphery is a dynamic category within the world economy and restructuring (p. 84) involves zones rising or sinking through the semi-periphery with consequences for their peoples and environments.

6.5 Politics and the world economy

In the world economy the extraction of economic surplus takes place primarily through expropriation via the market, but the use of political and military power to obtain surplus is not entirely eliminated. The logic of accumulation includes the logic of state-building, geopolitics and imperialism, and political processes extend beyond states since all social institutions have their politics. Wallerstein identifies four that are crucial to the operation of the world economy. **States** were introduced in the last chapter and are centres of formal power responsible for the laws that regulate all other institutions. **Peoples** are groupings of individuals that have cultural affinities. A cultural group may control a state, in which case it is defined as a nation, or it may constitute a minority within the state, when it may be referred to as a minority or an ethnic group. **Classes** are strata of the world's population based on economic criteria or their location within the mode of production. Finally, **households** are defined not by kin or cohabitation, but in terms of the pooling of income. Everyone is first and foremost a member of a household: that household is subject to the laws of a particular state, will have cultural affinities with a certain 'people' and will be economically located within a specific class. As these institutions interact, one upon another in diverse ways, they create and recreate the space–time matrix described above. They facilitate and constrain the behaviour of individuals by laws, rules, customs and norms, and shape what is and is not possible in terms of the social construction of nature and realization of sustainable development.

In the nineteenth century and first half of the twentieth century the leaders of European states used liberalism and nationalism to facilitate the rise of the world economy. Modern institutions erected in the name of freedom and progress became bastions behind which the propertied classes defended their interests, and old and newly constructed national identities were used as ideological tools to stress unity, emphasize threats from those who were clearly 'foreign' or 'different', and sustain imperialist adventures. **Nationalism** suggests that there are biologically or culturally distinctive communities of people, that each community should live under one political system independent of others, and have the right to demand equal standing in the world with others. It is often linked to imperialism and racism, but also finds expression in nationalist movements seeking independence from imperial rule. In the nineteenth century socialists who saw nation-states as props to capitalism advocated a form of **internationalism** that emphasized the international brotherhood of the working classes. Some sought to take the modern project in new directions that would give expression to a homocentric environmental ethic.

6.6 Socialism and homocentric environmental ethics _____

The industrial revolution crowded people into filthy, disease-ridden towns where they were subject to new economic and social rhythms. Changed social and environmental relations in the home and workplace alienated the working class from human and non-human nature but also fostered a new sense of solidarity as it recognized itself as the oppressed mass at the bottom of society. The basic forms of education required by the factory system also enabled the spread of the new political ideas of socialism, anarchism and communism, as did the popular forces unleashed by the French Revolution of 1789. These forces met a dismal end as monarchs reasserted their power, but there were further republican rebellions throughout Europe in 1830 and 1848.

The roots of **socialism** lie in Christian teaching and related ethical commitments to rationality (all people are born equal and gross inequality flouts the dictates of reason); solidarity ('united we stand, divided we fall'); and cooperation (as the basis of social organization). Socialism (Heater, 1974; Jones *et al.*, 1991) was developed by labour organizations and trade unions, movements to extend the franchise, and friendly societies based on mutual aid, that built on earlier utopian initiatives such as those of the Diggers and Levellers. The visions of Saint-Simon, Fourier and Owen, developed between 1750 and 1850, all emphasized the decentralization of power to small-scale communities, common ownership of resources, egalitarianism, respect for nature and harmony between town and country. Such thinkers followed Rousseau (1712–78) in arguing that democracy was only possible, and could only guarantee freedom, when people lived in small 'face-to-face' communities and fully joined in the process of law-making. Representative democracy as practised in the West was meaningless, making citizens free only for a few minutes every few years when they went to vote. Rousseau regarded people as essentially social in nature, and dismissed the idea of them living in a state of nature (p. 92), except as 'noble savages' without the use of language.

It was Karl Marx and Friedrich Engels who exposed the weaknesses of utopian socialism and anchored socialist politics in dialectical and historical materialism. They suggested that working people should realize their own futures, rather than follow blueprints provided by others, and that this would involve the popular control of the economy and state, not just the adoption of new values and ideas (Feuer, 1978). The growing contradictions of capitalism suggested that modern science and technology could produce a society of plenty, in which people's natural and species powers are developed to the full, and all contribute according to their ability and receive according to their need. The Communist Manifesto encouraged international socialism, but the second meeting of international socialists in 1889 ended in disagreement. Reformers believed that capitalism could be changed through representative democracy and the extension of human rights, while revolutionaries divided between communism and **anarchism**. Communists believed in strong party organization and the need for the socialist state to guide the transition to communism, while anarchists rejected the party and the state as authoritarian and undemocratic, preferring more direct forms of democracy based on decentralized and simpler ways of living.

Despite hardship, life for the majority of people in the West slowly improved. Trade unions improved people's working environments, private philanthropists began to build model settlements for their workers, and public health acts encouraged the new municipal authorities to provide such services as clean water. Faced with the threat of revolutionary socialism, the owners and managers of capital gradually accommodated those workers' and citizens' demands that were compatible with continued profitability. Rising material living standards that owed much to ecological imperialism were a key factor in establishing a dynamic social compromise that eventually took the form of social democracy. In Russia, where revolution did occur in 1917, state socialism proved to be undemocratic and its system of centralized economic planning exploited both human and non-human nature in ways that are described in the next chapter.

Socialism incorporates a **homocentric environmental ethic** or ecological humanism that balances immanence and transcendence, technocentrism and ecocentrism (p. 19) by regarding nature as an integral part of society that should be consciously planned and cared for in the interests of present and future human well-being. People should act as stewards of non-human nature, recognizing it as a source of economic, scientific, aesthetic, cultural and existence values that should be conserved and enhanced. Such an ethic finds support in Jewish and Christian teaching, is consistent with the application of science to environmental management, and can guide technology assessment. The visions of future socialist landscapes developed by William Morris and Peter Kropotkin (see Box 6.2) reflect such an ethic. Morris pioneered an ecological socialism, and its themes of useful work, simplicity, community, quality of life and the enjoyment of nature were rediscovered by the new environmental movement of the 1960s (Coote, 1995; Morton, 1979).

Box 6.2   Peter Kropotkin (1842–1921): anarchist and geographer

Kropotkin was born into an aristocratic Russian family and developed an interest in radical ideas while a young army officer. Service and travels in Siberia convinced him that exiles and colonists should govern themselves rather than be governed by remote authority, and that such self-government had a natural basis in the cooperative behaviour of the animals and humans that he observed. His reputation as a geographer was established by major scientific expeditions. After resigning his commission, and rejecting the prestigious post of Secretary to the Imperial Russian Geographical Society (on the grounds that he had no right to such a post while others around him starved), he spent several months with watchmakers in the Swiss Jura. Here his conversion to anarchism was completed and on returning to Russia, he became a revolutionary activist. Imprisoned as an agitator, he made a dramatic escape, fleeing to Britain where he worked as a writer. He continued to visit centres of anarchism in Europe, and after further imprisonment in France he led an increasingly quiet and scholarly life. He retained links with political activists and continued to travel, but fell out with many of his anarchist friends over his militarist attitudes towards Germany and the First World War. The Russian Revolution allowed him to return to his homeland, and he spent his final years protesting against the increasingly authoritarian

Figure 6.4   The anarchist geographer Peter Kropotkin
Source: National Museum of Labour History.

(continued)

(continued)
nature of Lenin's state. His funeral in Moscow was attended by 100,000 people.

In *Mutual Aid* (1902), Kropotkin argues that while competition between and among species does occur, by far the most important influence on animal and human evolution is the natural tendency for individuals to cooperate. He believed that mutual aid was a prehuman instinct or law of nature, and the book catalogues historical examples of mutual aid, for example animals hunting in packs and the shared use of resources by early tribes. Survival of the species depends, not upon competition between members of the species but rather upon cooperation within the species in order to withstand external threats.

In *Fields, Factories and Workshops Tomorrow*, Kropotkin imagines a future landscape or utopia shaped by anarchist principles. Settlements are decentralized and small-scale to allow self-government and enhance personal and collective fulfilment. Town and country differences are blurred to enable people to work in factories and fields, and enjoy fresh air and 'natural' beauty as well as cultural opportunities. Greater self-reliance at local, regional and national levels fosters people's control over their lives and undermines imperialism, but there is mutual aid between communities, regions and countries. Diversification means people and communities can produce a wide range of goods and services and are less dependent on others. Individuals do both mental and manual work, and labour-intensive production gives everybody useful work to do. Machines are used to enhance meaningful work, and workers and their families contribute according to their means and are rewarded according to their needs. By re-humanizing work Kropotkin's utopia puts people back in touch with human and non-human nature and so reduces alienation.

In 'What Geography Ought to Be' (an article published in 1885), Kropotkin advocates an anti-militarist, anti-imperialist and anti-capitalist education through geography that would examine issues from the point of view of the working class, foster social harmony and mutual aid, and involve students in the everyday life of their communities.

Source: Cook and Pepper, 1990.

6.7 The natural and social sciences

Socialism derived support from developments in organismic biology (p. 29) that taught that the whole is more than the sum of the parts, the parts are of equal value, and nature is constantly evolving and developing. Marx and Engels (Engels, 1950) offered dialectical materialism as a framework that could incorporate such ideas and unify the natural and social sciences, but the emerging social sciences proved reluctant to consider the biophysical world largely due to the influence of Malthus and Darwin. In his *Essay on Population*, published in 1798, Malthus suggested that there are insurmountable natural barriers to modern progress since population always increases faster than available food supplies. He derived his principle of population from empirical data, but by suggesting that increased poor relief would only allow population to press against food supplies more acutely, and that liberal and socialist notions of social progress contradict natural laws, Malthus was supporting the conservative status quo. History has disproved his argument, but it has continued to shape modern environmentalism (Harvey, 1974).

Darwin was a systemic thinker who stressed the close analogy of humans and animals. His *Origin of Species* (1859) claimed that evolution resulted from natural selection, or the survival of the fittest, with chance genetic variations that provide advantage being passed on and developed. Scarcity for resources (an idea borrowed from Malthus) was the basis of competition for survival, and Darwin's ideas can be seen both to reflect and justify liberal capitalism and imperialism. **Social Darwinism** applies evolution to socioeconomic and political affairs, suggesting that some human societies have competitive advantage and are superior to others, or that such phenomena as poverty and underdevelopment are due to biologically inherited characteristics. The notion of adaptation to environment was developed as environmental determinism (see below). Darwin's writing does contain passages that regard nature as a cooperative community in which niche specialization reduces competition, but it was left to radicals such as Kropotkin (Box 6.2) to expose the ideological use of his ideas.

The rise of modernity stimulated social scientists such as Marx, Weber and Durkheim to describe and explain the profound social changes taking place and suggest reasons why the promise of the Enlightenment was not being realized. Comte developed positivism in an attempt to discover objective laws operating in the conduct of whole societies, but it was largely geographers who concerned themselves with theories of the environment. These theories lagged behind mainstream social science and continued to reflect a false nature/society dualism.

6.8 Geography and imperialism

The emergence of geography as a modern academic discipline and its institutionalization within society is linked to the growth of trade, science and imperialism. Voyages of discovery, trade, military campaigns, and the administration of empires, required the expansion of cartographic and topographic knowledge, and a subject that would catalogue and order the vast amount of new information that Europeans were acquiring about the world. Geographical societies were established in Europe in the early nineteenth century (Berlin, 1828; Paris, 1821; London, 1830), but the first university chairs and departments came later in the 1870s. These societies and departments assisted governments with imperial policies that sought territorial acquisition, economic exploitation and the domination of local peoples, with geography stressing the superiority of white, Christian Europeans over the black, heathen races of Africa, Asia and Australia.

Alexander von Humboldt (1769–1859) and Karl Ritter (1779–1859) are generally considered to be the founding fathers of modern geography. In regarding the subject as a synthesizing science that should demonstrate the unity or holism of the natural world, Humboldt combined German romanticism and idealism with French materialism. His speculative and intuitive approach to an evolving, interconnected universe anticipated dialectical and systemic materialism, and is reflected in studies of landforms, climate, soils, vegetation and animal life that combine empirical method with a desire to reveal the principles that hold all nature together in an harmonious and functioning system. Humboldt's essentially ecological view of geography, underpinned by an agnostic and aesthetic appreciation of the order of the cosmos, contrasts with Ritter's historical and regional view that sought to reveal causal relations within the environment as evidence of God's grand design.

Under attack from geologists, biologists, sociologists and others, who refused to recognize geography as a scientific discipline, geographers built on the work of Varenius, Kant (pp. 92–3) Humboldt, Ritter and others, to focus on the region and the relations between people and their environment as organizing ideas (Pryce, 1977). The region proved to be a useful tool for classification and teaching, but geography remained largely descriptive with confusion about the nature and direction of the links between people and their environment and the relations between physical and human geography. Darwinian ideas were selectively assimilated by geographers who found the evolution of organisms through time and their adaptation to their environment more attractive than Darwin's central tenet of natural selection. The biological idea of adaptation to environment reinforced **environmental determinism** in geography: the view that the physical, natural or geographic environment (climate, soil, topography, vegetation, etc.) shapes or controls human activity and culture. Such determinism renders people passive by stressing their role in adapting to rather than shaping the environment, and serves as ideology by justifying poverty and racism in terms of 'difficult' environments which limit the production of wealth and/or the development of black people.

Environmental determinism was developed in Germany by Friedrich Ratzel (1844–1904), in America by Ellen Churchill Semple (1863–1932), and later in a more scientific form by Ellsworth Huntingdon (1876–1947) and Griffith Taylor (1880–1963). Ratzel emphasized the intimate relations that human communities appeared to enjoy with the land; suggested that human societies, like plant and animal communities, adjusted to their physical environment; and added a spatial dimension to Darwin's ideas by suggesting that nations had a natural tendency to expand and create empires unless constrained by stronger neighbours. Semple's determinism was 'crude' in that she lacked the data to support wild assertions and made no allowance for human free will. Huntingdon and Taylor attempted to quantify the effects that specific elements of the physical environment had on patterns of human activity. Huntingdon sought links between regions of invigorating climate and 'high' civilization, while Taylor developed an index of habitability based on environmental constraints that allowed him to assess the potential for settlement in Australia and support arguments against immigration from Asia. Taylor gave people and technology more prominent roles than earlier determinists, suggesting that the environment sets directions and that people act as traffic controllers, altering the rate but not the direction of progress.

Opposition to determinism came in the late nineteenth century from those who supported **possibilism**: the view that the physical environment is passive and that people are active agents at liberty to choose between a range of possibilities that the environment provides. French geographers, led by Paul Vidal de la Blache (1845–1918), were influenced by two sociologists: le Play, who urged an integrated social science built around the themes of work, family and place; and Durkheim who envisaged world society as an autonomous system possessing a morphology (formal pattern) and a physiology (lifestyles and behaviour). They suggested that the manner in which people react, respond and adjust to environmental resources is conditioned by traditional ways of life (the whole complex of attitudes, traditions, institutions and technology), with different groups developing entirely different ways of life within the same or similar physical environments. Their studies of the resulting cultural regions and landscapes were further developed in Britain by Herbert Fleure (1877–1969) and in America by Carl Sauer (1889–1975). Fleure outlined a scheme of world regions based on people's success in modifying environments, while Sauer interpreted environment as landscape, the product of culture evolving alongside biophysical phenomena.

Those nineteenth-century geographers who were concerned about environmental change and conservation worked from largely empirical and idealist foundations. Mary Summerville (1780–1872) recognized the new powers and responsibilities that industrial technology brought in its wake, while George Perkins Marsh (1801–82) urged environmental stewardship and management to conserve harmony and balance in the natural world. It was not until the 1950s that the limitations of environmental determinism were fully debated and exposed, and it was not until the 1970s and 1980s that radical nineteenth-century anarchist geographers (Box 6.2) were rediscovered and Marxist ideas about society and nature were incorporated into the discipline.

6.9 Further reading

We can learn much about an expanding world system by studying the exploitation of a single species: read Kurlansky (1999) on cod, or Williams (1988) on whales.

Hecht and Cockburn (1989) document the incorporation of the Amazon forest and its peoples into the world system and make the case for an 'ecology of justice'. Read about the defenders of the forest, including Chico Mendes, and update their account by finding out about current movements for land reform in Brazil. Alternatively read Moorehead (1987) on the invasion of the South Pacific between 1767 and 1840.

Goudie (1993) provides an introduction to geographical ideas about people's role in environmental change in his first chapter. Read this chapter, review the ideas, and list the main ways in which Goudie's approach to human impact on the rest of nature differs from the approach taken in this book.

6.9.1 Websites

Commonwealth Scientific and Industrial Research Organization, Australia: http://www.csiro.au/index.asp
Action for Aboriginal Rights: http://home.vicenet.net.au/~aar/

Chapter 7

Late modernity: from Cold War to Greenpeace

7.1 Introduction

The three decades following the end of the Second World War are remarkable for their reshaping of global patterns of control over resources, the rapid growth in mass production and consumption, the associated human impacts on environments and the emergence of unprecedented concerns about the finite and fragile nature of the Earth's environment. The first part of this chapter addresses the changing patterns of control over resources, explaining the ideological and geopolitical contexts which supported new ways of organizing an increasingly globalized system of production. The second part of the chapter investigates some of the environmental consequences of this rapid and uneven process of development through a comparative study of productivist rural environments in the UK and former Soviet Union. The third part of the chapter considers the rise of environmentalism as part of a more general challenge to the dominant ideologies of the late modern world system.

7.2 The global economic system: 1945–75

In 1945 the United States had emerged as the global hegemonic power. Germany and Japan were defeated nations, while Britain, France and the Soviet Union had suffered enormously from war. United States production dominated in terms of volume, technology and trade (Table 7.1). Economic and military domination were complemented by an ideological bundle that combined the political ideals of individualism, anti-imperialism and democracy

Table 7.1 Industrial production as percentage of world total

	1870	1913	1926–29	1936–38	1960	1980	1992
United States	23.3	35.8	42.2	32.2	36.1	28.7	23.5
Britain	31.8	14	9.4	9.2	5.6	4.3	4.7
Germany	13.2	15.7	11.6	10.7	7	9.3	7.4
France	10.3	6.4	6.6	4.5	4.3	5.4	4
Japan	–	1.2	2.5	3.5	3.6	9.9	15.1

Source: Porter and Sheppard, 1998, p. 409.

with the economics of liberalism and Keynesianism. Global military and economic security was seen to be dependent upon an international allegiance to this political–economic ideology. Communism was seen as a threat to such global allegiance. The anti-imperialist trappings of Marxism, with its emphasis on state-sponsored industrialization and collective entitlement to state welfare, formed an alternative ideology that might appeal to newly independent nations – an alternative that was strongly at odds with capitalist expansion.

In the United States and Germany, industrial production had fallen to 53 per cent of its 1929 levels by the early 1930s (Wallace, 1990). However, the overriding logic of a capitalist system is an expansionary tendency, and recession is generally broken by industrial restructuring organized around spatial, technological and managerial transformations. In the three decades after the war, capitalist expansion was built upon technical innovations in electronics, chemicals, communications, agriculture and energy. However, profitability was also enhanced through geographical expansion, and the further globalization of chains of commodity transactions. Success in a capitalist system is measured by the ability to continually accumulate capital (Wallace, 1990; Hopkins and Wallerstein, 1996) and United States industrialists were, on the whole, in the strongest position to achieve this. However, there were obstacles to pursuing this ambition. Firstly, in order to realize the ambition of capitalist expansion, a state cannot operate unilaterally: other states have to be willing to play the game and this requires the legitimation of global processes of capital accumulation. In previous societies, the legitimation of capital accumulation has been gained through force (as in the British colonization of Australia) and through recourse to a divine hierarchy (as in premodern societies). However, the new world order sought legitimation through 'voluntary' subscription to the belief that an expanded arena for production and trade would ultimately benefit all participants. The United States had to sponsor a liberal international order by projecting its own values of mass production, free trade and private ownership (Corbridge and Agnew, 1995). Such an international order was to be challenged as it became increasingly apparent that the accumulation of capital on a global scale was a highly uneven process, developing and reinforcing wealth gaps and also resulting in uneven ecological burdens.

A second obstacle to the United States' ambitions came in the form of the war-torn economies of Europe and Japan. Europe had lost much of its industrial and colonial incomes – profits that had previously stimulated markets for United States exports (Reifer and Sidler, 1996). The solution to Europe and Japan's 'dollar shortage' was the Marshall Plan – a massive programme of loans and grants distributed for the rebuilding of war-torn economies, quickly restoring productive capacities and leading to increased trade with the United States. For the Soviet Union, the Marshall Plan, along with the NATO military alliance and the Mutual Security Treaty (with Japan), represented a form of informal imperialism which increased the threat of United States influence at its borders: it was an 'empire by invitation' (O'Tuathail et al., 1998). For the Third World, the focus of United States investments largely reinforced the core–periphery divide.

A third obstacle to expansion was the colonial trading blocs which circumvented free trade and free-market exchange. In response, the United States somewhat hesitantly promoted independence, although the drive for independence within the colonies was often irresistible in any case and the United States was more concerned with the political destiny of emerging states. The creation of multilateral institutions of global governance provided a locus for the promotion of liberal economics. The 1944 **Bretton Woods** talks resulted in agreement to principles of free markets and the benefits which expanded trade networks could bring. Such principles were institutionalized in the International Monetary Fund (IMF), which functioned to maintain the conditions for efficient market exchange, the International Bank for Reconstruction and Development (now the World Bank), which could aid integration into global markets through loan provisions, and the General Agreement on Tariffs and Trade (now the World Trade Organization, WTO), a forum for the promotion of free trade.

7.3 Cold War geopolitical order: 1945–75

While the United States sought to export its own set of ideals, so too did the Soviets, casting themselves as the true anti-imperialists. The crystallization of the **Cold War**'s binary opposition of East and West defined United States domestic and foreign policy. On the domestic front, extreme anti-communist sentiments (McCarthyism) could be employed to undermine labour movements and 'dissident' political opinions, while on the global map, such sentiments could be called upon to gain Congressional support for political, economic and military interventions in distant environments. Foreign policy was in some respects contradictory. On the one hand, America presented itself as an anti-imperial power, promoting political independence within the periphery. On the other hand, there was a concern that independence left a political vacuum that was a breeding ground for nationalism or pro-Soviet allegiance. Such fears led to ambiguous support for decolonization. Express anti-colonial stances, such as that which followed the 1956 re-invasion of Egypt by France, Britain and Israel, were motivated by the big polarized picture rather than by local sensitivity. America's stance in the 'Suez Crisis' was not principally founded on Egypt's right to self-determination, nor on the right of a sovereign state to control its own environmental resources (i.e. the Suez Canal). Rather, it was founded on the belief that such flagrant military intervention would provide ideological purchase for anti-Western sympathies, thereby playing into the hands of Moscow (Crockatt, 1995).

The 'war' between the two superpowers was mostly fought over Third World territories. While the publicly stated motives offered an unproblematic fight against evil, it is clear that the United States was not always committed to democracy and often helped to secure seats of power for ruthless dictators (Box 7.1). Nor were United States actions always anti-imperialist, for they reflect American fears about the consequences of unfettered self-determination. It is

Figure 7.1   British troops digging in along the Suez canal, 1956, following battle at Port Said
Source: Hulton Getty Picture Collection.

Box 7.1   The CIA and the control of resources

Iran, 1953

The CIA's role in Iran typifies its dilemma regarding self-determination and sovereign control over environmental resources. In 1951 Dr Mohammed Mossadegh was democratically elected as prime minister. He soon angered the British by nationalizing Iran's main oil corporation, the British-owned Anglo-Iranian Oil Company. In 1953, following British requests, the United States employed the CIA to support a coup against Mossadegh, eventually replacing him with the Shah. Economic embargo and CIA covert action were decisive in Mossadegh's early demise and in the outcomes, which included a new international consortium to distribute Iranian oil, leaving the United States with an economic and political foothold in the Middle East.

Guatemala, 1954

In 1951 Jacobo Arbenz was elected president of Guatemala with an overwhelming majority. Arbenz was openly leftist in his policies, promising land reforms and soon confronting American economic interests. Belief in his communist tendencies was reinforced by his nationalization of 225,000 acres of unused land, owned by the United Fruit Company (UFCO). UFCO was a major player, controlling Guatemala's biggest export, bananas, and owning

(continued)

(continued)

an astonishing 42 per cent of the nation's land. The World Bank cut off all loans to Guatemala, the United States began a trade embargo and UFCO paid for a massive publicity campaign to persuade the United States government to intervene directly. In 1954 a CIA-assisted coup took place with the help of American planes and American-trained rebels. General Castillo Armas was installed, UFCO got its land back, the land reform programme was shelved and United States aid started to flow.

Congo (Zaïre), 1960

The Congo gained independence from Belgium in June 1960 and Patrice Lumumba was democratically elected as the first prime minister. The United States had branded Lumumba a communist and aided in his assassination and replacement by President Mobutu, a 'friendly' dictator who aided the plunder of Zaïre's natural resources for the next 37 years. In 1963, when rebels loyal to Lumumba threatened to overthrow Mobutu, they were crushed by Belgian troops carried in United States planes.

Indonesia, 1965

In 1948 George Kennan (head of policy planning in Washington) described activities in Indonesia as the United States' most important struggle with the Kremlin. In that year the British had temporarily overthrown the fledgling independent government led by President Sukarno, while the Dutch were seeking a return to power and a return to the important stream of capital accumulation which this colony had yielded before the war (estimated at 20 per cent of Dutch GNP). Sukarno's nationalism, combined with his active *non-alignment* in the Cold War, eventually led to a CIA-supported attempted coup. In 1965 the CIA's renewed intervention assisted General Suharto into power and within weeks one of the bloodiest persecutions of the twentieth century took place, with up to a million alleged 'communists' murdered. Suharto ruled for 33 years, allowing the violent oppression of native peoples to make way for the exploitation of rich forest and oil wealth by domestic and international interests.

Sources: Zepezauer, 1994; Crockatt, 1995; Hopkins and Wallerstein, 1996; Chomsky, 1996; Reifer and Sidler, 1996.

often hard to dissociate the ideological high ground (saving the free world from the communist menace) from the vested economic interests which flowed from ensured access to commodities. Control over the circuits of capital is never far from the equation, adding substance to Wallerstein's World System Analysis (Chapter 6).

7.4 Trade, production and resources in the capitalist world _____

Trade involves the flow of goods and services over space and helps to determine where goods are produced, where they are consumed and the division of labour within the global production process (Cho, 1995). In a world without international trade, all goods must be produced and consumed within state borders. If Norwegians had a passion for mangos they would have to invest in large heated greenhouses and produce them domestically, at great cost. The ideology of free trade is partly concerned with removing the need for such inefficient production, ensuring that production of commodities is concentrated in locations which are best equipped for the purpose. In nineteenth-century economic thinking, especially in the work of David Ricardo, places which are 'best equipped' for particular production processes are those which inherit the most appropriate forces of production. These include climate, mineral resources, energy, labour, capital, technology and management skills. Ricardo argued that there was mutual benefit to be gained from different regions in the world specializing in the production of goods for which they had a **comparative advantage**. The pursuit of comparative advantage featured heavily in post-1945 economic policy and was institutionalized within the Bretton Woods organizations. Free trade and production specialization could lead to an increase in consumer choice, more efficient use of resources, an increase in total global production and the opportunity for economic growth in the periphery. While reliance on overseas production might be deemed a threat to domestic security, subscription to United States' economic ideals could also bring the security of its military umbrella.

7.4.1 Production in the periphery

Despite the rhetoric of mutual gain, comparative advantage led to a marked **international division of labour** and a broadening of wealth disparities between core and periphery. Prior to the late 1970s, comparative advantage tended to lead to the concentration of manufacturing industries in the economic core (and the Soviet Union), while raw material production was concentrated in the periphery. The core nations exported manufactured goods because they had an abundance of technology, skilled labour and capital, while peripheral countries exported raw materials because they had an abundance of natural resources and cheap labour. In principle, an advantage was an advantage, regardless of the particulars of the product. Malaysia's comparative advantage in producing rubber could lead to the profitable accumulation of capital in just the same way as Germany's comparative advantage in producing motor vehicles. However, while this may have been temporarily true, it had become clear by the early 1970s that comparative advantage was leading to a global hierarchy of production. While Ricardo's theory could (imperfectly) explain where production was located, it could not explain the prices of the products on world markets (Cho, 1995). Relative market prices are denoted by **terms of trade**, the relationship between the prices of a nation's exports and imports.

Most significantly, the terms of trade deteriorated for most peripheral economies because the market price of raw materials fell relative to the market price for manufactured goods. The consequences include the need to extract more raw materials in order to maintain imports of manufactured goods (and the servicing of debts) and a system of capital accumulation that enriches the most industrialized nations at the expense of those with a comparative 'advantage' in raw materials. By 1974, Third World governments were lobbying the United Nations for a **New International Economic Order**, a package of demands in which trade reforms were prominent (Todaro, 1996; Hadjor, 1993; Ghatak, 1995). Of course, the supposed 'advantage' of specializing in certain peripheral processes was as much contingent on historical imposition as an evaluation of postwar assets. In Malaysia, for example, rubber was not grown because of its comparative advantage for local people, but because it was imposed by colonial rulers and entrepreneurs who could profit from it themselves. The same is true of cocoa in Ghana, tea in India and coffee in Brazil (Porter and Sheppard, 1998).

7.5 Environments in the United Kingdom

Following the disruption of world wars and depression, industrial capitalism was fuelled by the new American world order, the technological revolution, the birth of mass consumerism and political consensus arising from rising living standards. In the 'white heat of the technological revolution' (Harold Wilson), people had 'never had it so good' (Harold MacMillan speaking in 1959) and were fulfilled in 'keeping up with the Joneses'. But it was also an age in which industrial capitalism was to reveal more clearly a long-harboured nemesis: the apparent contradiction between continuous growth and the limits of nature. The postwar consensus was built upon a mixed economy in which the liberalization of trade and capitalist relations of production were tempered by Keynesian management of the national economy. The state managed capitalism in a way that produced a marriage between consumerist individualism, capitalist entrepreneurship and welfare provision (Slater, 1997), but there was little emphasis on maintaining the biophysical conditions of production.

7.5.1 Postwar rural environments

In the nineteenth century British agricultural policy had been influenced by an advantage in industrial production, a belief in free trade and confidence in the empire as a source of primary production. The repeal of the Corn Laws in 1846 was a move towards a free market in agricultural goods and the opening of the Suez Canal in 1869 reduced the costs of imports from India, Australia and New Zealand. The government maintained a belief that cheap imported food would help to suppress urban wages and maintain industrial competitiveness (Marsden *et al.*, 1993; Corbridge and Agnew, 1995). By 1900, only 20 per cent of the UK's population lived in rural areas and only 8 per cent of the workforce was engaged in agriculture. Britain imported 75 per cent of its wheat and 50 per cent of its meat (Marsden *et al.*, 1993). By the 1930s, 70 per cent of New Zealand's entire exports consisted of foodstuffs destined for Britain and, by the outbreak of the Second World War, two-thirds of all food was imported (Marsden *et al.*, 1993). The 1942 Scott Report noted the desperate state of rural communities who had been virtually abandoned by urban and empire-oriented government priorities. Between 1926 and 1936, 884,000 acres of agricultural land was lost, mostly to urban development, but nearly one-quarter of this simply reverted to moorland, abandoned by farmers who could no longer farm the land profitably (Clapp, 1994).

The response to agricultural decline has been described by Shoard (1987) as the 'third agricultural revolution', following the neolithic revolution (see Chapter 4) and agricultural enclosures of the eighteenth century (Chapter 5). Postwar agriculture underwent massive transformation, stimulated by the decline of empire, the security fears following the German U-boat campaign, new technologies and government interventions. UK government and later European Community support for agriculture led to a rapid industrialization of farming, a coalition of farmers with industry that transformed rural areas into modern, productivist environments. The capitalization of agriculture involved investment in plant and machinery, increased use of chemical and energy inputs, the concentration of larger-scale, specialized farm units in the hands of fewer farmers, greater food-processing, marketing and transportation, regional specialization and changing cropping patterns. Capitalization has led to the uneven development of agriculture, with economies of scale dividing farmers into the 'haves' and 'have nots' (Robinson, 1990). The average farm size grew from 28.1 ha in 1935 to 75.9 ha in 1985 (Robinson, 1990) while the number of farm holdings fell from 524,905 to 254,198 between 1939 and 1987 (Blunden and Curry, 1988). Between 1950 and 1980, the agricultural labour-force halved (Grigg, 1995). The total value of farm expenditure on inputs such as animal feed, labour, machinery and fertilizers rose from £737 million in 1950 to £9,758 million in 1988 (Bowler, 1991).

Restructuring of farm holdings, together with increased inputs has been successful in achieving higher yields and production (Table 7.2). However, there have been notable failures of postwar agricultural production: the uneven distribution of benefits, the fall in rural employment, overproduction leading to the notorious 'food mountains' and huge burdens on national expenditure: by 1983, over 60 per cent of the European Community's budget went on supporting agriculture. Such lavish expenditure paid little heed to medium- and long-term environmental costs. The positioning of farming within global circuits of competition has required highly intensive methods. With the relatively high labour costs (compared with producers in the periphery), farmers have sought to gain competitiveness through intensive transformations of environments. Some of the main costs associated with this are: loss of wetlands, species-rich grasslands, moorland and hedgerows; extensive loss of flora and fauna; nitrate pollution of drinking water; pesticide pollution; eutrophication of rivers and lakes; soil erosion and degradation; loss of landscape amenity.

Table 7.2 Production and yield in selected UK crops

Crop		1939	1944	1958	1968	1978	% change 1944–78
Wheat	P	1,672	3,189	2,755	3,469	6,613	+107
	Y	2	2.2	2.7	3.5	5.3	+141
Barley	P	906	1,780	3,221	8,270	9,848	+453
	Y	2.2	2.2	2.9	3.4	4.2	+91
Oats	P	2,035	3,001	2,172	1,224	706	−76
	Y	2	2	2.4	3.2	3.9	+95
Sugarbeet	P	3,586	3,320	5,835	7,119	6,382	+92
	Y	25.8	19.3	33.1	38.1	31.7	+120
Potatoes	P	5,302	9,243	5,646	6,872	7,331	−21
(maincrop)	Y	18.6	16.1	17.1	24.6	34.2	+112

Source: Adapted from Bowler, 1991, p. 90.
P = Production in 000 tonnes.
Y = Yield in tonnes per hectare.

From the 1970s geographers contributed to increasing criticism of productivist policies, eventually heralding a shift towards post-productivism, with greater emphasis on goals relating to rural livelihoods, environmental conservation and countryside recreation.

7.5.2 Mass consumption

The demand for food is relatively inelastic because there are limits to how much a person can eat. Population growth and changing diets can account for some of the UK's increased agricultural production during the postwar decades, but the majority has found markets due to import substitution and exports. However, the same cannot be said about manufacturing production. In the core states, annual growth in **GNP** averaged 5.1 per cent between 1950 and 1973 (Ikeda, 1996) and the great majority of manufactured goods remained within the core. Much of this growth therefore fuelled (and was fuelled by) rising consumption within the core. Indeed, it was in the 1950s and 1960s that mass, **Fordist** consumption became of central importance to modern capitalism (Bocock, 1993). **Mass consumption** had taken off rather earlier in the United States, with Henry Ford's 'Model T' motors being assembled in Detroit from 1910. Cultures of affluent consumption had also existed in Europe in earlier times, but the 1950s and 1960s saw a new era, distinguished by the fact that advertisers were targeting working-class people. By 1972, 93 per cent of UK households owned a television set, 66 per cent a washing machine, 52 per cent a car and 43 per cent a telephone (*Social Trends*, 1975).

The postwar years saw the rise of the supermarket, electricity, consumer durables, brand images, teenagers, 45 rpm records, frozen fish fingers, car

Figure 7.2   Ford Motor Company plant, Dagenham, 1961
Source: Hulton Getty Picture Collection.

Table 7.3   Cars per 1,000 population, UK: 1953–92

	1953	1961	1966	1971	1976	1977	1981	1984	1992
UK	57	116	181	225	253	260	317	343	380
West Germany	22	95	179	247	308	326	385	412	464
France	47	135	210	261	300	315	348	360	419
Netherlands	18	55	121	212	269	283	323	332	373
Sweden	60	175	241	291	351	346	348	nd	413
Italy	13	50	125	209	284	290	322	359	491
United States	288	345	400	427	510	530	536	540	578
Japan	1	8	28	102	160	180	209	227	313

Source: *Social Trends*, HMSO, 1980; 1987; 1995.

ownership and holidays in the sun (Tables 7.3 and 7.4). As the postwar economic 'boom' enabled ever greater numbers of workers to satisfy an ever-growing list of marketed 'wants', the new world order became hard to criticize (Slater, 1997). However, critical thinkers did confront the myth of progress and freedom engendered by rising consumption. Firstly, some Marxist analysts such as Baudrillard (1970; transl. 1998) viewed the new consumption as a form of social control, not breeding self-expression but, instead, hooking people into conformist treadmills in which hollow contentment was achieved through the meeting of manipulated wants with standardized goods. In fewer words,

Table 7.4 Holidays abroad and holiday entitlement for UK residents

	1951	1961	1966	1971	1978
Holidays taken abroad (millions)	1.5	4	6	7	9
Holidays abroad to Spain (% of holidays abroad)	19	21	27	29	30
Annual holiday entitlement, full-time manual workers (%):					
two weeks		97	63	28	0
two to three weeks		2	33	5	1
three weeks		1	4	63	17
three to four weeks		0	0	4	47
four weeks and over		0	0	0	35

Source: *Social Trends*, HMSO, 1980 (British National Travel Survey, Dept of Employment).

Table 7.5 Distribution of wealth in the UK: 1962–77

% wealth owned by:	1962*	1966	1971	1974	1977
Wealthiest 1%	36.7	33	30.5	22.5	24
Wealthiest 5%	62.2	55.7	51.8	43.1	46.4
Wealthiest 10%	75.4	68.7	65.1	57.5	61.1
Wealthiest 25%	93.7	86.9	86.5	83.6	83.9
Wealthiest 50%	100	96.5	97.2	92.9	95
Least wealthy 50%	–	3.5	2.8	7.1	5

Source: Inland Revenue.
*1962 data assumes no wealth for least wealthy 50 per cent.

consumers were dupes. Secondly, general improvements in living standards did little to change the unequal ownership of wealth, particularly for the poorest 50 per cent of society (Table 7.5). Thirdly, the perpetuation of new wants, premised on capitalism's structural requirement for ever-expanding production for profit, seemed to contradict the reality of finite resources. Fourthly, the need for increased profit for producers potentially contradicted the need for consumers to buy more: profit could be increased through labour-shedding technology, reduced wages for labourers or cheaper environmental inputs. The first two threatened to dampen market demand, while the latter appeared to defy the economics of scarcity.

7.6 Environments in the former Soviet Union

This book has analysed relationships between society and nature that have evolved within different societies with different forms of social and economic organization. Under the general categorization of 'modes of production', hunter-gatherer, tribal agricultural, feudal and capitalist societies have been featured,

Soviet communist economic relations	Western capitalist economic relations
capital-intensive production (post 1928)	capital-intensive production
social ownership (state ownership)	private ownership
production for use (for distribution, not exchange)	production for profit (for sale on market)
no markets (some exceptions); state fixes prices (determined by value of labour)	market determines prices/values
no competition between producers no competition for capital	competition between corporations competition for capital
wages paid by state	waged labour for wealth-owning classes
surplus extraction by state no individual accumulation of wealth	surplus extraction by class of owners personal profit
centralized (command) economy, run by state	mixed economy – state planning and free market
protected from world economy	integrated into world economy (some protection)

Figure 7.3   Economic organization in the former Soviet Union

demonstrating how relationships between society and nature are inextricably linked with relationships within societies. In Karl Marx's historical analysis of social progress, he concluded that advanced capitalist societies had been developing the conditions under which a further transformation would occur, to a communist mode of production. Surveying the state of the working class in the mid nineteenth century, Marx and Engels not only saw the injustice that existed, but also envisaged an inevitable resolution to such inhumanity between people. Essentially, they envisaged the working classes becoming politicized and exercising irresistible force to instigate a socialist revolution wherein the wealth-owning classes would be abolished and ownership of the means of production would accrue collectively to the working class. While the transition to this new mode of production would resolve the antagonistic and exploitative relations that existed between people, it would also resolve the antagonistic relationship between humans and nature. The abolition of private property in favour of collective ownership of the means of production would reconcile the chasm between people and nature because nature (as part of the means of production) would no longer be a medium of workers' exploitation. In his *Economic and Philosophical Manuscripts* of 1844, Marx (1975) wrote of this as superseding people's estrangement (or alienation) from nature. To simplify a rather curious metaphysical argument, **alienation from nature** is superseded when nature becomes 'for everybody' rather than 'for the elite'. Working with nature would then become a free, non-exploitative activity in which individuals are reconciled with both nature and their own humanity. In this early work, Marx saw the resolution of social problems as being inseparable from the resolution of society–nature problems because: 'To say that man's physical and mental life is linked to nature simply means that nature is linked to itself, for man is part of nature' (p. 328).

The progression from advanced capitalism to communism never happened. In the former Soviet Union, the October Revolution of 1917 was not premised on advanced industrial capitalism and, arguably, it was not premised on the political awakening of the majority class. However, it did lead to political and economic characteristics that are distinct from capitalism (Figure 7.3), although all would agree that it hardly resembled a form of socialism espoused by radical thinkers in Europe and the United States. Workers have been exploited by the bureaucracy, there has been a lack of popular participation in planning and management, political and cultural expression has been suppressed, production has been inefficient, food has often been scarce and, in some sectors, environmental damage has been even greater than in the West.

7.6.1 Nature and Soviet Society

The Soviet state sought to modernize living standards through a massive and swift industrialization of both manufacturing and agriculture. Owning all land and industry, the state could accumulate surplus and direct it towards massive infrastructural investments, resulting in production units far greater than those found in the West. By the mid 1970s, the average industrial enterprise in the United States employed 48 people, compared with a figure of 565 in the Soviet Union (Lane, 1985, p. 9). However, while the economy grew rapidly, we are now well aware of the apparent disregard for both people and environment. Attitudes within the state bureaucracy show no reconciliation with nature, only a powerful desire for its conquest. Behind the façade of stringent environmental standards a remarkably cavalier attitude revealed itself.

So where did such attitudes stem from? Some commentators (e.g. Wolfson, 1992; Ziegler, 1987) point to the impact of size on perception. The shear enormity of the territory (16 per cent of the world's land surface), with some of the world's richest reserves of agricultural lands, forests, minerals and fuels, gave rise to perceptions of infinite wealth. A second observation is that this was very much a modern, enlightenment, revolution, inspired by the potential achievements of science when applied within the context of a rationally planned economy. Indeed the faith in official science (that sanctioned by the state's ideology) was perhaps even greater than that experienced in the West and was certainly less exposed to independent scrutiny. There was a faith in the judgement of the bureaucracy: that transformations of nature would lead to predictable and controllable outcomes and that any undesired outcomes were transitional, manageable anomalies (Ziegler, 1987). Such **technocentric** environmental managerialism was equally evident in the West, although the Soviet state displayed a particular bravado. Conquering nature through massive transformations was a way of symbolizing the power of the Soviet system, not only to its own people but to the rest of the world. Marxist histories regarded industrialization as both inevitable and progressive; achieving the maximization of the forces of production was a necessary part of the transition to communism and the premature nature of the October Revolution required urgent transition. Under Stalinism, obstacles to such historical progress were to be overcome, whether they came in the form of Orthodox religious consciousness or physical constraints. Prior to Stalin's death, in 1953, the conquest of nature was closely linked to the profound exploitation of humans. The *gulag* system of prison workers was a system of enslavement in which KGB (secret police) agents were given quotas for arrests. Quotas had to be exceeded by whatever means necessary and untold millions of mostly innocent people were set to work digging canals and working mines – labour for which the state could not afford wages.

Underpinning Stalin's – and subsequently Khrushchev's – abuse of nature was a crude interpretation of materialist philosophy. Stalin believed that material relations of production were the only relevant historical forces. As such, it was folly (and anti-revolutionary) to suggest that nature could or should play a role in determining social activities. Nature could perhaps influence the pace of economic progress, but it could not determine its characteristics. As such, nature was conceived as a passive object, lacking any powers of constraint over humans. This is crude materialism because it ignores the sophistication and complexity which dialectical relationships entail. In a dialectical relationship between society and nature there is more of a flavour of co-evolution, of inseparable realms that continuously influence each other, resolving contradictions through mutual feedback and adaptation. In Soviet environments, feedback was itself suppressed and this significantly undermined the potential for effective environmental management.

7.6.2 Central planning of rural environments

A dual system of agriculture was established with 'collective' and 'state' farms. On collective farms, land is owned by the state, but is managed by the workers who collectively own the tools of production and, to an extent, the produce. The state strongly influences production plans and sets procurement targets which itemize the quantities and prices of products that the collective must sell to the state (Lane, 1985). The collective makes decisions regarding the use of income, including re-investments into the business and payments into welfare and culture funds. State farms are mostly larger in size and all the means of production are owned by the state; workers receive a minimum wage plus production bonuses. In addition, many farmers are allocated private plots which can be used for subsistence production and, since 1932, for sale to *kolkhuz* markets (Libert, 1995). The variety of agricultural environments, from arctic tundra to sub-tropical Central Asia, is too huge to detail here. Suffice it to say that the range of crops grown varies over space and that regional specialisms have emerged, ranging from reindeer-herding to rice-farming.

Within a largely self-sufficient economy, cheaply produced agricultural surplus is a condition for urban and industrial expansion. Thus, Soviet agricultural environments were very much subservient to industrialization programmes, and state appropriation of crops was often draconian. Rural environments functioned to deliver cheap food and raw materials for distribution to non-agricultural sectors. In the 1950s this function was largely pursued

Figure 7.4 The Virgin Lands programme, Soviet Union
Source: Adapted from Libert, 1995 (Figure 2.1).

through the geographical expansion of agriculture, i.e. through extensification. In the 1960s intensification became more significant with investment in inputs such as drainage, irrigation, fertilizers, pesticides and herbicides. Extensification is epitomized by Khrushchev's Virgin Lands programme which involved the ploughing of 40 million hectares of new arable land, mostly in the grassy 'steppes' of Kazakhstan (Figure 7.4). Such grand-scale transformation of nature paid too little attention to local conditions (a common problem with central command) resulting in the ploughing of inappropriate lands, some of which were completely unsuitable for agriculture (Libert, 1995). Widespread soil erosion and dust storms resulted in the 1960s and were only partly dealt with through tree-planting and the development of ploughless tillage techniques.

Intensification of agriculture in the 1960s and 1970s is strongly associated with hydraulic management. Perhaps the grandest scheme of all was the Siberian rivers' diversion project, with the intention to take waters from the Ob and Irtysh rivers (flowing north into the Arctic Ocean) and divert them 1,500 km south into southern Russia and Central Asia. This project has been on and off since the 1940s and now seems to be buried. However, the irrigation systems that did take place were equally grand in totality. Major rivers were transformed into irrigation facilities, regularly dammed, with huge reservoirs and vast networks of connecting irrigation canals. In the 1970s, 600,000–700,000 hectares of land was newly irrigated each year (Libert, 1995). While irrigation has advantages, such as increased reliability and yields of crops, it has also had enormous environmental consequences, the most infamous case being the tragedy of the Aral Sea (Box 7.2).

Box 7.2   The Aral Sea: dossier on disaster

Location: The Aral Sea is in Kazakhstan and Uzbekistan, in what was Soviet Central Asia (Figure 7.4).

Size: In 1960 the surface area was approximately 68,000 sq. km, making it the fourth largest lake in the world. It is now the eighth largest lake in the world. Some 80 per cent of the water volume has been lost and 58 per cent of the surface area.

Inflows: The lake is fed by the rivers Amu-Dar'ja and Syr-Dar'ja. Originally, the combined annual discharge was 60 sq. km. It is now less than 10 km and the sea is being lost through evaporation.

Irrigation: River water is mostly extracted for the irrigation of cotton and rice crops. The region now has eight million hectares of irrigated land and the majority of people rely on this for their livelihoods. Uzbekistan became the world's third largest cotton producer.

Fishing: Some 'coastal fishing villages' are now 100 km or more from water. In the 1940s annual fish catches of more than 40,000 tonnes were recorded. Since 1980 no fish have been caught. All but two indigenous fish species have disappeared.

Table 7.6   The desiccation of the Aral Sea

Year	Area (sq. km)	Volume (cu. km)	Sea-level (m)	Salinity (g/l)
1960	68,000	1040	53	10
1976	55,700	763	48	14
1985	45,713	468	41.5	23
1990	38,817	282	38.5	30
1995	35,374	248	37	40
1998	28,687	181	34.8	45
2010 (predicted)	21,058	124	32.4	70

Sources: Micklin, 1992; German Remote Sensing Data Centre, 1999. The Aral Sea homepage http://www.dfd.dlr.de/app/land/aralsee/chronology.html

(continued)

(continued)
Impacts

The loss of the Aral Sea is a disaster. Valuable ecosystems have been destroyed, crops are threatened, the fishing industry has collapsed and human health has been tragically affected. Wind erosion of the desiccated basin leads to salt and sandstorms over hundreds of kilometres. Crops are covered, soils are salinated, and rates of throat cancer and lung and eye diseases have risen. Infant mortality has soared and high levels of pollutants in remaining waters have led to the rise of diseases such as typhoid and cholera. Pesticide concentrations have also risen as remaining river water is largely drained from surrounding fields having been extracted upstream. Even the climate is thought to have been affected: albedo has increased seven times and reflected radiation three times, leading to warmer summers, colder winters and a shorter growing season.

Solutions

Apart from 'more irrigation' through the Siberian rivers' diversion project (p. 127), the only real solution lies in massive diversification of the economy, away from irrigation-dependent farming. One estimate is that halving the irrigated area of Uzbekistan would stabilize the sea – but six million people would lose their livelihoods. Despite global awareness over many years, and many plans and commitments, the fact is that nothing has been done that comes close to addressing the social basis of this problem.

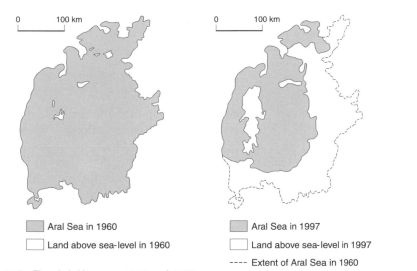

Figure 7.5   The shrinking sea, 1960 and 1997
Source: Adapted from German Remote Sensing Data, 1997; Bathymetry of the Aral Sea.

(continued)

(continued)

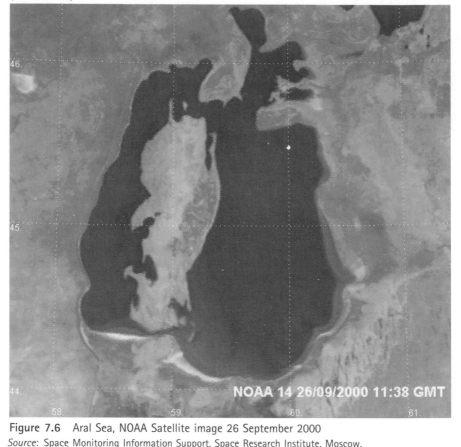

Figure 7.6   Aral Sea, NOAA Satellite image 26 September 2000
Source: Space Monitoring Information Support, Space Research Institute, Moscow.

Ever since experiences in ancient Mesopotamia (Chapter 4), it has been known that successful irrigation requires effective drainage. Without this, surface water and rising water tables lead to problems of waterlogging and salinization. Such problems have been widespread in drier parts of the former Soviet Union, partly driven by the fact that success for party officials was measured in terms of area irrigated rather than the efficiency of drainage and water transfer. Adding to the problems of this target-driven development was the uniformity of the commitment. Central plans were often too grand in scale to be sensitive to local variations in cultural and environmental conditions and this was exacerbated by the lack of information flows between the local and the central. On the one hand, local knowledge was not always fed back to planners and, on the other hand, central knowledge of negative environmental impacts was suppressed. Officially, plans were always going very well. At root were the attitudes discussed above, and particularly the faith in officially recognized science to prevail in the conquest of nature for social progress. In common with capitalist societies, science and technology cannot be considered as entirely neutral and objective; it is partly a product of prevailing modes of organization and dominant ideologies (Chapters 8 and 12).

7.7 New social movements and the environment

It is perhaps no surprise that this period of rapid industrialization led to renewed concerns about the environment. In Europe and America, these concerns became prominent towards the end of the 1960s, in the periphery towards the end of the 1970s and, in the former Soviet Union, following *Glasnost* in the late 1980s. Despite earlier movements, the new wave of environmental organizations is generally characterized as something unprecedented and often associated with the concept of **new social movements** (NSMs). These movements are historically distinct because they transcend class-based social organization. The 'old' social movements refer to working-class labour movements that challenged both working conditions and the system of economic organization. NSMs can be equally radical, in the sense that they challenge dominant ideologies and seek to change the world. The women's movement is an exemplar, successfully challenging patriarchal organization and ushering in some of the most significant social reforms of modern times. The extent to which such new movements are a progressive continuation from 'old' class-based historical struggle, has been much debated (e.g. Yearly, 1994) but it is noteworthy that many NSMs have been closely associated with leftist politics.

NSMs may or may not evolve out of long-running historical agendas, but they are certainly responsive to contemporaneous social and economic climates. The Cold War, and particularly the Vietnam War, did much to spawn the peace movement in the United States and Europe, while arms escalation and nuclear accidents fuelled different wings of the anti-nuclear lobby. The environmental movement was a response to loss of habitat and species, resource economics, local and transnational pollution, the treatment of animals by society and the erosion of green spaces. Rachel Carson's (1962) seminal work *Silent Spring* referred to the impact of pesticides on United States' wildlife, explaining the absence of birdsong on spring mornings. Why such issues should have led to an international movement in the late 1960s requires some further examination, as does the question of why such a movement should provide a challenge to industrial capitalist ideologies. Referring only to Western movements, one theory is that rising affluence among a broader base of the population gave rise to a material climate in which new values could emerge. New values underpinning parts of the environmental movement were clearly post-materialist, recognizing that rising material standards of living could only satisfy certain sets of needs. The treadmill of affluent consumption was destroying environments and did little to satisfy higher needs such as spiritual fulfilment and self-expression. The 'hippy' and commune movements commonly

Figure 7.7 Women protest against US nuclear missile base at Greenham Common, 1981
Source: Hulton Getty Picture Collection.

expressed elements of such a post-material value system. Post-materialism was also associated with post-industrialism (Doyle and McEachern, 1998). As the industrial model of social progress was progressively exposed for its failures to bring harmony within society and, most pertinently, between society and natural systems, the Enlightenment ideology was ripe for challenge. The internationalism of environmental concern was encouraged by the transnational nature of the issues involved and by developments in the media. Images became important symbols of concern, typified by the sense of fragility conveyed by the first pictures of Earth from space. The love affair between Greenpeace and the media really took off in 1976 when images of the clubbing of baby seals provoked immediate international outrage and a massive surge in membership.

The scientific basis of early environmentalism largely focused on issues of population and resources. Fossil fuels, agricultural land, water and other vital natural resources were forecast to be running out. For some (such as Schumacher, 1973), scarcity was the prompt for questioning values and priorities within the core, in a combined search for better lifestyles and better environments. Economic growth was even forecast to bring lowering rather than improving living standards. For others (such as Goldsmith *et al.*, 1972) the response was more **ecocentric**, focusing on how non-human nature could be protected, with less attention paid to advancing human welfare. Uniting both approaches was a challenge to economic growth, the fundamental requirement of capitalism. Environmentalism also contained a politics of self-interest that involved looking for the enemy without. In such a construction, concern for the future lay less with the big consumers in the core and more with the burgeoning populations in the periphery (e.g. Ehrlich, 1972). While many social movements wanted greater empowerment for people, the politics of scarcity, based on **neo-Malthusian** catastrophism, often sided against this. Garret Hardin's (1968) 'The Tragedy of the Commons' identified the roots of over-exploitation in systems of local and communal management of natural resources. This influential article concluded that resource tenure and management should be further concentrated in the hands of the state and private enterprises (see Chapter 10).

In the periphery, environmental movements were generally inseparable from issues of local livelihood and were certainly not born out of affluent consumption. The Chipko movement in the Indian Himalayas typifies this. Its origins lay in local people's struggle to protect their own livelihoods by challenging the commercial and state forestry practices that excluded them from effective control over their own natural resources. It was first a social movement, second a women's movement and arguably only third an environmental movement. But it is the very inseparability of these three strands, bound up in the practical necessity for local people to participate in the management of their own environmental resources, which was the real strength of this movement. The fact that this peaceful campaign was successful in securing Indira Ghandi's intervention, a recognition that social and ecological problems had common roots in unjust systems of landownership and control, remains an inspiration to many non-governmental organizations in India and elsewhere. It provides a useful antithesis to Hardin's more conservative analysis.

In the former Soviet Union the increasing release of information about the state of the environment, along with the legalization of independent organizations, fuelled a massive, though fragmented, environmental movement that challenged both the cohesion of the Soviet Union and its system of state industrialization. Here, environmentalism was often associated with nationalist struggles for independence, rooted in anger at the ways in which the Moscow bureaucracy had abused regional environments. Goldman (1992) reported that virtually any new industrial proposal was being met with mass protest, to the extent that it was even difficult to replace old, polluting plants with newer ones.

7.8 Geography and social context

The study of geography has not been immune from social contexts and in Chapter 6 it was seen that the focus of geography in the core was influenced by colonial interests. In the 1950s geography became more scientifically orientated, adopting the increasingly outdated tenets of logical positivism (Chapter 2). This shift was in part a practical response to a need to compete within a university system in which 'harder' scientific disciplines were attracting greater esteem. But David Harvey, himself a proponent of scientific methods in the 1960s, also saw this shift as related to Cold War politics (Livingstone, 1992). In the United States in particular, he suggests that geographers had to hide behind the apparent neutrality of science to avoid being accused of communist sympathies. This is not entirely unlike the position of Soviet geographers who had to avoid being branded anti-communist. From the late 1960s this 'scientific revolution' was itself out of synch with prevailing social contexts. For example, a discipline that could not speak out about social injustice and its connection to degraded environments could not honestly claim to be a discipline committed to studying the relationship between people and their non-human environments. David Harvey was himself to become one of the leaders of a new, socially committed expression of geographical knowledge and activity. He was not only committed to exposing inequalities but also to analysing the roots of injustice through a Marxist analysis of capitalism's structures of economic organization. Geographers had to abstract from observable phenomena and identify the structural roots that underpinned their dynamics. Ultimately, such analyses sought both to interpret the world and to change it, very much in the spirit of the time.

7.9 Further reading

Another book in this series, Klaus Dodds' (2000) *Geopolitics in a Changing World* provides an excellent introduction to the area that includes a chapter on environmental geopolitics.

For a highly informative analysis of relations between core and periphery, see Porter and Sheppard's (1998) *A World of Difference*, especially Part III.

There is much to be said about New Social Movements that we have not been able to cover. Start with Doyle and McEachern's (1998) *Environment and Politics*.

7.9.1 Websites

For further information and updates about the Aral Sea:
http://www.dfd.dlr.de/app/land/aralsee/ (Aral Sea homepage, German Remote Sensing Data)
http://smis.iki.rss.ru (for the latest satellite images)

Chapter 8

Late modern society and global warming

8.1 Introduction

The publication of the Brundtland Report, *Our Common Future*, by the World Commission on Environment and Development (WCED, 1987) popularized the conceptual link between developmental and environmental needs. The call for 'sustainable development' included the argument that many environmental problems could only be effectively addressed by tackling both social and ecological foundations. Solutions to global environmental problems had to go hand-in-hand with political efforts to tackle inequality of resource distribution and to achieve both intergenerational and intragenerational equity. This report was based on research conducted by an international team and reflected social scientific approaches to the environment that had been developing since the early 1970s. It was also reported to be very influential and, to this day, remains one of the most-cited books on the environment. However, in this study of late modern environmental management, focusing mainly on global warming, it is suggested that the social-developmental dimensions of environmental issues continue to be sidelined and that modernist, technocentric solution frameworks predominate.

8.2 Economy, environment and science

Early environmentalism was characterized by conceptions of scarcity that owed much to classical economic perceptions of the link between finite resources and future human well-being. Economic recession in the mid 1970s was met by economic restructuring along **neoliberal** lines: the reduction in state intervention in the economy and a move from mixed economies to more market-based regulation. Neoliberalism is closely associated with **neoclassical economics**, a more optimistic body of economic theory that rejects the classical notion that economic growth is necessarily subject to natural constraints. For neoclassical theorists, the market can, in principle, provide the mechanism for resolving issues of scarcity and pollution. For example, when a resource becomes scarce, its price will begin to rise, thus stimulating resource-augmenting measures such as exploration, substitution, recycling and other technological innovations (e.g. Simon, 1998). Furthermore, the market is viewed as a medium for democratic participation in environmental management because individuals are able to express their environmental (and other) preferences whenever they buy or sell on the market. Both buying goods (shopping, investing) and selling labour (working) involves the democratic exercise of what are dubbed 'dollar votes'. When such 'votes' are cast in ways that protect environmental resources (for example by purchasing organic vegetables or investing in ethical unit trusts) it is known as **green consumerism**. The aggregate of these individually expressed preferences amounts to a social mandate that businesses are ultimately accountable to.

Prophecies of scarcity have often proved over-alarmist, with science and technology achieving what neoclassical economists have forecast. However, the market, in tandem with green/ethical consumerism, has not proved to be an adequate environmental manager in late modern times. Scarcity is still exemplified by the woeful inability to prevent loss of biodiversity and by the shameful failure to distribute resources to the 800 million people who cannot even afford adequate nutrition. Indeed, left to the market, food is exported out of areas with high levels of malnutrition, to areas where people can afford to pay for it. Furthermore, as fears about scarcity (temporarily) recede, concerns over pollution have raced up the agenda. Employing a neoclassical analytical framework, pollution arises from rectifiable market failures that result from **externalities**. These are entities that are external to economic markets and therefore not managed by them. For example, the emission of carbon dioxide into the atmosphere represents an externality: in the past, there has been no specifically designated cost associated with such emissions, despite the fact that future costs will almost certainly accrue to society. Lack of a market has effectively meant that this form of pollution has been 'free' and where something is free it is generally over-used.

In dominant constructions of environmental problems, the market's past failure to prevent pollution is acknowledged. However, the deeper problem of the market's failure to prevent increasing wealth disparity is generally denied. In ignoring this structural link between environment and development, the solutions to crises are often in the form of 'more markets' rather than less, and can be classed as **accommodatory**. In such solutions, the existing relations of production can accommodate new forms of market intervention such as 'green' taxes, emissions standards and tradable quotas. In the UK, for example, there are plans for new cars to be taxed according to their carbon dioxide emissions from late 2000, rather than a flat rate or a rate differentiated by engine capacity. Transnational environmental problems arising from market failures have increasingly been dealt with through new forms of regional and global governance, such as the European Union and the United Nations. The depletion of the ozone layer, acid rain and climate change are problems that are as global in scope as the economic system in which they are generated. They cannot therefore be resolved unilaterally and require international consensus and cooperation.

8.2.1 Markets and knowledge

The market's ability to protect the environment is partly determined by access to knowledge. Governments and consumers are unable to act in 'green' ways if they don't understand the consequences of their decisions. For example, DDT (an organochloride pesticide developed in the 1930s) was widely believed to be an unproblematic scientific discovery that could prevent crop loss and protect millions from the misery of vector-borne diseases such as malaria. CFCs were seen as another triumph of science, providing cheap and safe coolants that could help in the mass ownership of refrigerators and consequent improvements in food hygiene. Atmospheric sulphur dioxide was viewed as a relatively unproblematic emission from power generation. All turned out to be environmental menaces and it must be concluded that their 'market externality' was more to do with lack of knowledge than lack of markets *per se*.

Transformations of nature have always led to both intended and unintended outcomes. Intended outcomes are the ones we want, while unintended outcomes are only rarely desirable. In the past, the tendency has been to maintain a belief in science's ability to predict the outcomes of human activities. Unintended consequences have been viewed as resulting from underdeveloped science, rather than a fundamental impossibility of understanding the complexities of socio-biophysical systems. However, in the late modern era, leading scientists are telling us that the world is a far more complex place than was envisaged in early modern times. Simple mechanical laws do not always operate and scientists now deal in 'indeterminacy' as much as they deal in firm predictions. The implications for environmental management are (or should be) revolutionary. In the absence of perfect prediction, uncertainty has to be managed by building flexibility and resilience into planning processes and into environments themselves. For example, if one accepts that a major river system cannot be managed in a fully predictable way (its behaviour is indeterminate) there is a need to develop resilience by way of reducing vulnerability to the unexpected. For example, rice farmers in Bangladesh have traditionally grown very tall varieties that can stand above floodwaters. Overall, it is generally fair to say that developing resilience and flexibility requires social rather than market-based responses (although the two are not inseparable). At the global level, one of the key responses to uncertainty is the development of strong and effective global institutions that have the capacity to respond flexibly and swiftly to emergent complications. This is sometimes referred to as the **new institutionalism**, a sort of rebirth of the political idealism of the interwar years, as crystallized in the (failed) League of Nations (see O'Riordan and Jordan, 1999). At local levels, resilience might best be developed through the democratization of environments and the development of local capacity to respond to emerging challenges. Resolving these potentially contradictory objectives of centralization (globalization of institutional reach) and devolution (localization of environmental control) may well be one of the biggest challenges of this century.

The following case-study of global warming brings in questions of externalities (greenhouse gases), uncertainty (is global warming real and what will its impacts be?), solution frameworks (technocentric and social) and 'new' institutions (the United Nations and the Intergovernmental Panel on Climate Change).

8.3 Global warming

The scientific community has demonstrated a sensitivity to uncertainty, reminding policy-makers that their work is dealing in best estimates, that their models of global climate are inexact and that there are Earth processes for which there is simply too little understanding to advise upon. This has proved hugely frustrating for those seeking urgent action in response to global warming. Policy-makers are relatively inexperienced at managing uncertainty and often make the inappropriate response of 'awaiting further proof'. Within a neoclassical economic ideology, policy-makers want precise predictions that enable them to put a monetary cost on future impacts. The method is to evaluate costs of current preventative action in relation to the benefits that will accrue from them in terms of averting future costs.

Global warming is often referred to as the 'enhanced greenhouse effect'. Greenhouse gases such as water vapour, carbon dioxide and methane act like a blanket by allowing more radiation into the Earth's atmosphere than is allowed out again. This imbalance occurs because these gases are more efficient at absorbing long-wave, reflected radiation than they are at absorbing incoming, short-wave radiation.

Without the greenhouse effect, the planet would be an estimated 33 °C colder and unsuitable for life. However, atmospheric concentrations of greenhouse gases are not constant and are subject to both naturally occurring and anthropogenic (human-induced) change. For example, the rapid increase in the burning of fossil fuels over the last couple of centuries has released considerable quantities of carbon dioxide into the atmosphere. The precise pathways of CO_2 accumulation are complex and uncertain, but rising atmospheric concentrations are beyond doubt (see Box 8.1).

The Intergovernmental Panel on Climate Change (IPCC) have produced a 'best estimate' that surface temperatures will increase by a further 2 °C by 2100 (range 1–3.5 °C) and that thermal expansion and ice-melt will lead to a further rise in sea-levels of about 50 cm (range 15–95 cm) (IPCC, 1996c). However, there are many complications that mean that warming could occur much faster (or slower), that significant spatial and temporal variation will inevitably occur, and that precipitation patterns will also change. Uncertainties include:

- The effect of rising temperatures on cloud formations is not well understood and could lead to both positive feedback (knock-on effects that intensify the warming process) and negative feedback. On balance, the former is considered likely to be predominant (Brown, 1996).
- The role of ice sheets and frozen tundra. Melting of the main Antarctic ice sheets is considered highly unlikely, but the melting of mountain glaciers, frozen tundra and sea ice shelves is likely. This can itself lead to positive feedback systems: ice melt will leave a more absorptive surface that will reflect less heat back into the atmosphere, and the melting of tundra would release huge quantities of trapped methane.

Box 8.1   Evidence of human–induced climate change

We can make some generalizations about what would constitute sufficient evidence to prompt policy responses. Firstly, evidence is required of anthropogenic contributions to greenhouse gas concentrations. Secondly, there is a need to demonstrate that the climate is indeed changing and, thirdly, a need to demonstrate a *probable* link between the two.

1. Evidence of anthropogenic contribution to greenhouse gas concentrations

Carbon dioxide (CO_2) is largely added to the atmosphere through the burning of fossil fuels, but is also produced by burning and decomposing woody biomass and industrial processes such as cement production. Pre-industrial concentrations of atmospheric CO_2 were approximately 280 parts per million (ppm), and had risen to 363 ppm by 1996 (World Resources Institute, 1998), currently rising at an average annual rate of 3 ppm. Anthropogenic sources of Methane (CH_4) largely arise from land-use operations, including cattle and sheep production, coal and oil extraction, rice production and landfilling of organic waste. Methane is found in lesser concentrations, measured in parts per billion (ppb), but has a much higher radiative forcing effect than CO_2. Concentrations have risen from pre-industrial levels of 700 to 1670 ppb (World Resources Institute, 1998). Nitrous oxide (N_2O) arises from biomass burning, fertilizer use and fossil fuel combustion. Concentrations have risen from 285 to 310 ppb. CFCs are also greenhouse gases but their significance is now generally played down owing to the fact that they also break down another greenhouse gas (ozone) and their concentrations are destined to fall following international action. Their replacements (such as HFCs) will rise in concentration and contribute to radiative forcing in the next century (IPCC, 1996b). Overall, the correlation between increased emissions and rising atmospheric concentrations is convincing and has led the IPCC to conclude that 'these trends can be attributed largely to human activities' (IPPC, 1996a). Furthermore, CO_2 is not destroyed in the atmosphere, it is added and subsequently redistributed through exchange with other CO_2 sinks such as oceans and forests. The path-ways of exchange are complex but we do know that many of them take a long time: for example, the exchange of atmospheric CO_2 into vegetation, and subsequently into peat and then coal and oil, is a process counted in millennia. Such complexity makes it impossible to give a single time-frame for the life of CO_2 in the atmosphere, but, taken as a qualitative estimate, a figure of 50–200 years can be entertained (IPCC, 1997a; Brown, 1996).

2. Evidence for recent climate change:

• Global mean surface air temperature has risen by between 0.3 and 0.7 °C this century.

(continued)

(continued)

- Recent decades have witnessed some of the hottest years since records began in 1860. In 1998 global surface temperatures hit an all-time high, exceeding by 0.2 °C the previous record from 1995 (WMO, 1999; NASA, 1999).
- The ten warmest years on record have all occurred since 1983 (WMO, 1999).
- 1998 was the twentieth consecutive year with above-normal global surface temperatures (WMO, 1999).
- The five-year mean temperature has risen by approximately 0.5 °C since 1975, representing the most rapid period of global warming on record (NASA, 1999).

3. Evidence of a link between human activity and recent global warming.

The above evidence may be enough to convince you that human-induced climate change is a reality. But climate scientists have a much harder job than this. Complexities emerge because any human-induced warming is mixed up with naturally occurring cycles and events. Scientists have to disentangle the anthropogenic from the natural, and it is here that sceptics have the greatest opportunity to raise doubts. Four examples will serve to illustrate 'natural' complications:

- Surface temperatures are known to have changed between glacial and interglacial conditions over the last 160,000 years, demonstrating long duration cycles of warmer and colder periods. The last 10,000 years has been an interglacial phase, associated with 'natural' warming resulting from the Earth's orbit of the Sun.
- Shorter cycles, measured in centuries, also occur. For example, Europe enjoyed a particularly warm period in the twelfth and thirteenth centuries, a 'mini ice age' in the seventeenth century and, currently, another warm period.
- Even shorter cycles occur, such as the warming effect of the El Nino Southern Oscillation, an event that should certainly lead to some caution when interpreting temperature data for 1998.
- Isolated events, such as the 1991 eruption of Mount Pinatuba, affect global climate. Fine particles (aerosols) emitted into the stratosphere act as a barrier to solar radiation and lead to a cooling effect that can last from months to years.

One source of historical data comes from ice cores. In the Antarctic, snowfall is sparse and ice develops over long periods of time, trapping and compressing air within it. Russian scientists have taken an ice core that contains ice accumulated over the last 160,000 years, enabling concentrations of atmospheric gases to be measured over this period. Identifying temperature changes over such periods is difficult, having to rely on 'proxy' evidence such as the levels of certain oxygen isotopes present in ice and the distribution of pollen types (Houghton, 1994; Huntley, 1990). Two things can be gleaned from

(continued)

(continued)

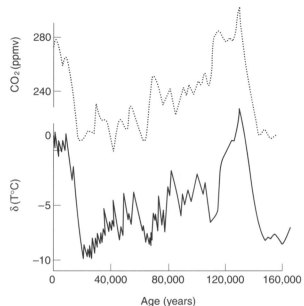

Figure 8.1 Atmospheric carbon dioxide (top) and temperature change (bottom), as inferred from the Vostok (Antartica) ice core over the past 160,000 years to the last interglacial
Note: The temperature change δ°T here is the temperature relating to the Antarctic Ocean water evaporating and then condensing as snow.
Source: Cowie, 1998, p. 35.

such evidence: firstly, that there appears to be a relationship between atmospheric carbon dioxide concentrations and global temperature and, secondly, that the industrial period has coincided with an increasing concentration of such gases.

In the attempt to distinguish between anthropogenic and natural influences, scientists have to deal in the language of statistics, i.e. they look for probabilities rather than certainties. Here, most studies have 'detected a significant change and show that the observed warming trend is unlikely to be entirely natural in origin' (IPCC, 1996b). The IPCC is typically cautious in its choice of words, but research into the attribution of causality has been intense. IPCC modelling of the hypothesized relationship between human activities and climate change has become more and more sophisticated as they incorporate increasing numbers of variables. Importantly, their models (run through computer simulations) are beginning to correspond to what has actually been observed in the world. In fact, the correspondence is now sufficiently tight to persuade most scientists that there is only a very low probability that it is pure coincidence. Another way of stating this is to say that there is a very high probability that the hypothesized link between human activity and global warming is correct.

- The behaviour of ocean currents play a huge role in determining climate regimes. For example, the UK is exceptionally warm for its latitude; it is heated by a warm ocean current that progresses northwards up the Atlantic to the point at which its salt concentration and falling temperature makes it fall below less dense waters. It then turns on itself and, deep in the ocean, retraces its passage (Figure 8.2). A possible scenario is that increasing precipitation over the North Atlantic will lower the salt concentrations of this warm current, delaying its sinking and resulting in dramatic regional changes in temperature (Houghton, 1994).
- Aerosol particles, such as sulphur emissions from power stations and ships, reflect radiation away from the Earth's atmosphere and also contribute to cloud formation. They therefore have a cooling effect that partly offsets warming. Current models generally assume constant emissions of such coolants, but reductions are likely because of their contribution to other problems such as acid rain and respiratory disease.

Uncertainty, and the very real possibility of rapid, non-linear changes, should bode for caution rather than delayed reaction. We cannot predict exactly what 'surprises' will occur, but we can predict that some are likely to happen. Even without sudden, non-linear changes, the predicted pace of warming is beyond the limits of comfortable adaptation. Many parts of the world are extremely vulnerable to climate change and sea-level change, with parts of Africa, for example, relying on crops that are growing at the extremes of their tolerance to both temperature and moisture regimes. Warming is predicted to occur at too fast a rate for ecosystems to adapt and re-establish themselves (IPCC, 1997b). Forests, for example, cannot rapidly migrate to colder latitudes or higher altitudes.

8.3.1 Science, modernity and ideology

In the words of the IPCC Chair, scientific reports should be 'policy relevant, but policy neutral' (Watson, 1999, p. 4). In previous chapters, it has been seen that science, whether in thirteenth-century England or twentieth-century Russia, has often been inseparable from political and religious ideology and therefore far from neutral. Despite this, modernity is characterized by a belief in the objectivity and neutrality of science, believing that scientists put a mirror up to nature and 'see it as it is'. Such a belief is founded on profound philosophical assumptions. It requires an ontology that accepts the existence of an objective physical reality and an epistemology that holds that humans can directly experience the workings of this reality without any interfering bias from their own mental baggage. Such a modern, positivist philosophy leads to simplistic assumptions about the relationship between knowledge and power: the IPCC and other scientists observe the workings of physical nature and report this 'neutral' knowledge to policy-makers. Policy-makers then evaluate the financial costs and benefits of alternative responses to this information, and make (rational) decisions about interventions. Knowledge is therefore seen to inform power, but not to be influenced by it: the relationship is assumed to be unidirectional and deterministic, rather than dialectical.

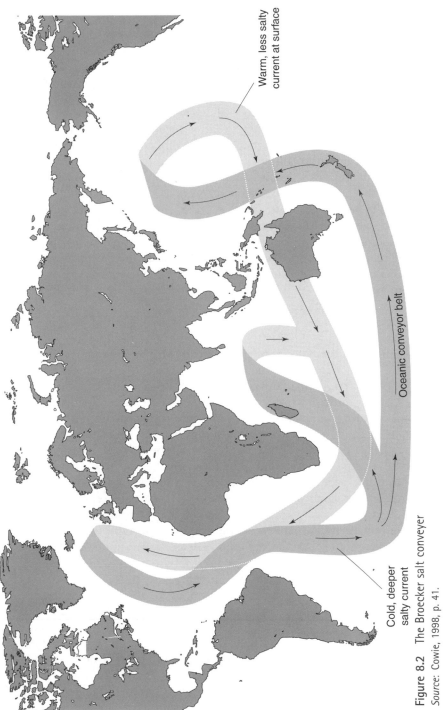

Warm, less salty current at surface

Oceanic conveyor belt

Cold, deeper salty current

Figure 8.2 The Broecker salt conveyer
Source: Cowie, 1998, p. 41.

In arguing that knowledge and power are dialectically related (that, to an extent, they are mutually constituting), we do not argue that science is incapable of representing nature in a useful manner: indeed anyone in Europe or Asia who turned out to view the August 1999 solar eclipse should have been impressed by the predictive capabilities of modern science. More to the point, without scientific observation and analysis we wouldn't even be aware of the greenhouse effect. However, it is necessary for social scientists to be cautious about the objectivity and neutrality of science. We need to be aware of the ideological and institutional contexts that permeate people's minds and within which scientific discovery and articulation are embedded. If we were not aware of this, we could not explain how two people can survey similar evidence and arrive at conflicting conclusions about whether the Earth is round or flat, or whether global warming is exaggerated or not. Our knowledge of real, physical processes needs to be related to the contingent circumstances within which the knowledge is constructed, not least because this can contribute to a better understanding of how people respond to issues such as global warming. Such contexts are, of course, partly geographical, relating to the physical–cultural places in which people live and the intellectual and emotional spaces to which they are networked. We will see below, for example, that people living in low-lying islands will tend to accept more readily the hazards of rising sea-levels than mainland dwellers who make their living out of fossil fuel-related industries.

A number of recent social scientific works on the environment have been critical of the modernist approach to the issue of climate change. First of all, they point to a continued faith in the predictive capacity of science. Indeterminacy is frequently portrayed as a temporary pathology (Wynne, 1994) that can be cured, for example, through more powerful supercomputers. In asserting such a creed, two important assumptions are reproduced within the climate change discourse. Firstly, faith in a determinist science is an assertion of the physicality of nature and lacks sensitivity to its social context. Nature and society are portrayed as independent entities, acting upon each other as if across some imaginary divide. To acknowledge social scientists' claims that nature is in part socially constructed would be to admit that it couldn't be fully understood by objective scientific enquiry. Secondly, in this philosophical separation of the physical from the social, an important but largely neglected decision is made regarding the ordering of environmental management. The 'correct' process is determined to be (a) identify the pure, objective science of nature; (b) proceed to the impacts of humans on this 'natural' process; and (c) predict the consequent impacts of a changed nature on humans (see MacNaghten and Urry, 1998). The IPCC itself embodies the prioritization of the physical and technological over the social and the resultant primacy of physical sciences. Even within its own ranks, Working Group I (science) is considered more solid and influential than Working Group III (socio-economic dimensions). Further, within the IPCC's institutional structure, the social dimensions of climate change are only considered relevant in relation to

Table 8.1 Nationality of IPCC authors

Report	United States	India	China	UK
IPCC 1990 Working Group I	110	5	8	62
IPCC 1996 Working Group I	158	3	5	61
IPCC 1996 Working Group II	154	14	8	24
IPCC 1996 Working Group III	30	7	2	5

Source: Agarwal, Narain and Sharma, 1999, p. 31.

the responses to the *impacts* of climate change on society. Social relations are excluded from considerations of the *causes* of climate change.

This might appear a rather indulgent and introspective discussion. However, to overlook the social contexts of atmospheric pollution (and science itself) places severe limits on the range of solutions considered. The decision not to analyse the social *causes* of environmental degradation (such as the roots of uneven development or capitalism's requirement for perpetual growth) is not policy-neutral. It delimits the boundaries of officially acceptable debate (Litfin, 1994) and constructs problems in ways that prevent open discussion of the massive social challenges that need addressing (Wynne, 1994; Visvanathan, 1991). As Jäger and O'Riordan (1996) suggest, a modernist scientific approach ignores the 'brutal politics' that lead to uneven development within the world system. We should therefore be aware that the primacy of science, and within that the decisions that scientists make regarding what is relevant enquiry, represents a deep epistemological commitment that is by no means neutral. In particular, we should be aware that this commitment tends to rule out serious debate about causes and responsibility (Shiva, 1993), a convenient consequence because the major industrialized nations are responsible for 84 per cent of historical (1800–1990) emissions from fossil fuel-burning (von Molthe and Rahman, 1996). These nations also dominate the IPCC panels (Table 8.1).

While deep-rooted epistemological assumptions serve to frame dominant ways of thinking about and responding to environmental problems, there are also more obvious ways in which values and interests provide a context for knowledge. Firstly, it is not controversial to state that the generation of knowledge serves defined interests and progress occurs fastest where there is political or industrial patronage. In a highly generalized way: what is funded determines what is researched which determines what is discovered. Secondly, data-collection might be relatively value-free but the articulation and interpretation of findings is often less so. For example, the Washington-based World Resource Institute (WRI) managed to articulate greenhouse gas emissions data in such a way as to conclude that 'Asia . . . is the largest contributor among the major regions of the world' (1990, p. 15). This choice of data manipulation drew swift criticism from Delhi's Centre for Science and the Environment (see Agarwal and Narain, 1991; Agarwal, Narain and Sharma, 1999)

Table 8.2 CO_2 equivalent emissions: 1995

Nation	Total CO_2 equivalent emissions (000 tonnes)	Per capita emissions (tonnes)
United States	5,468,564	20.5
UK	542,140	9.3
China	3,192,484	2.7
India	908,734	1
Bangladesh	20,932	0.2
Mozambique	993	0.1

Source: World Resources Institute, 1998.

and might be contrasted with the choice of articulation in Table 8.2. Note that articulation involves choices over the level of aggregation: emissions per person, per nation or per continent (Jäger and O'Riordan, 1996).

In addition to the level of aggregation presented, New Delhi's Centre for Science and Environment (CSE) felt that the WRI had selected a biased methodology for calculating a country's net emissions of greenhouse gases. In calculating net emissions, the WRI take into account the amount of greenhouse gases that can be absorbed and stored in environmental sinks, preventing accumulation in the atmosphere. Their error, according to the CSE, was to allocate sinks in proportion to emissions. Thus, as the United States was responsible for 26 per cent of CO_2 emissions, it was allocated 26 per cent of global sinks, greatly reducing the calculated net emissions. As an alternative, the CSE suggested that sinks might be allocated on a per capita basis; thus, with less than 5 per cent of the world's population, the United States would be allocated only a 5 per cent share in the world's sinks. Clearly, how one chooses to manipulate and interpret data makes a great difference. Using the same data, the WRI formula assigned 10.3 per cent of net global greenhouse gas emissions to India and China; the CSE formula assigned just 0.6 per cent (Agarwal, Narain and Sharma, 1999).

The interpretation of climate-change data can be even more value-laden. For example, in costing the impacts of future warming (part of the neo-classical process) it is deemed 'policy neutral' to value the life of a human being in the North up to ten times higher than one in the South (Jäger and O'Riordan, 1996). In terms of policy, such a costing mechanism can lead to the conclusion that more money should be invested in preventing the delete-rious effects of climate change in wealthy countries than in poor countries.

While climate-change science cannot be entirely objective, a critical realist philosophy does not throw us into the void of relativism, i.e. the belief that there is no 'true knowledge' because it is all relative to the situated interests of dominant agents. Common sense, based on personal experience of the practical uses of science, should inform us that scientific knowledge is often successful in revealing the 'real' physical workings of nature. For example, to deny this would be to deny a certainty that the Earth orbits around the Sun. However, a realist epistemology also recognizes that there are structural contexts within which society's relationship with nature is constructed and revealed. These contexts partly emerge from existing material relations of production and their expression in the dominant interests of the elite and nation-states. However, as will be clarified in Chapter 9, such structural interests are not impervious to other involved agents: the 'lay public' cannot be considered as hapless pawns, befuddled and manipulated into ways of thinking that suit the world's rich and powerful. As Beck (1992) has argued, there is growing resistance to dominant scientific worldviews.

8.3.2 The UN Framework Convention on Climate Change

Half a dozen supercomputers churn out simulations of the planet's possible climatic trajectories under differing concentrations of atmospheric greenhouse gases. Politicians can select their preferred trajectory, based on perceptions of economic costs and benefits. If they want to stabilize atmospheric CO_2 concentrations immediately, they can opt for an immediate reduction in emissions of 50–70 per cent, followed by further reductions (IPPC, 1996c). If they are happy with a few degrees more warming, they can attempt to stabilize greenhouse gas concentrations at double their pre-industrial level involving medium-term stabilization of emissions and long-term reductions (Figure 8.3). If highly risk-tolerant, and willing to discount costs to future generations heavily, they can allow 'business as usual' for the time being. Some of the following points about the political contexts within which scientific knowledge is debated will help to understand why, as of late 2000, world leaders have failed to agree on an acceptable trajectory to work towards.

In working towards an agreed future (an 'acceptable' level of expected warming), a long negotiation process led to a UN Framework Convention on Climate Change (UNFCCC), signed at the 1992 'Earth Summit'. The stated objective is

> [the] stabilization of greenhouse gas concentrations in the atmosphere at a level that would prevent dangerous anthropogenic interference with the climate system. Such a level should be achieved within a time-frame sufficient to allow ecosystems to adapt naturally to climate change, to ensure that food production is not threatened and to enable economic development to proceed in a sustainable manner (Article 2, UNFCCC).

The December 1997 Kyoto Protocol was signed by 39 industrialized nations. It did not clarify the 'concentrations' or 'time-frames' referred to above, but did introduce a binding agreement to reduce emissions of six greenhouse gases by an average of 5.2 per cent, from1990 levels, by 2008–12 (Table 8.3). Subsequent progress has been slow.

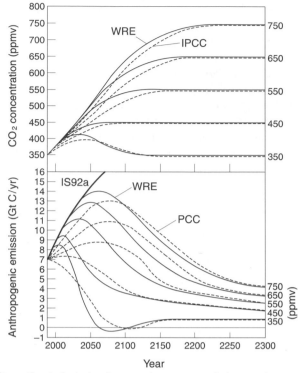

Figure 8.3 Alternative trajectories for greenhouse gas emissions and concentrations
IPCC = Intergovernmental Panel on Climate Change. IS92a is a 'business as usual' scenario.
All others involve degrees of future reductions in emissions. Note that for both sets of predictions, stabilizing atmospheric concentrations at 350 ppmv (lowest line on graphs) will require
massive reductions in anthropogenic emissions (lower graph), bringing net emissions down to
zero during this century. Even this will involve a temporary rise in concentrations (upper
graph). This graphical representation of options provides a powerful decision-making tool
but its message is so stark that governments have not been able to commit to a preferred
trajectory. When you have read Box 8.1, you might consider which you think is the most
sensible trajectory and what kind of changes this would require of the world economy.
Source: Wigley, Richels and Edmonds (1996), p. 240.

Table 8.3 Agreed cuts in six greenhouse gases under Kyoto Protocol*

European Union	8%
Canada, Hungary, Japan, Poland	8%
United States	7%
Russia, Ukraine, New Zealand	0%
Norway	+1%
Australia	+8%
Iceland	+10%

* Percentage reductions on 1990 levels to be achieved by 2008–12.

8.4 Geopolitics and international institutions _____

In an ideal world, actors participating in global environmental institutions such as the UNFCCC will act in the common interest; after all, it is in everybody's interest to resolve transnational environmental problems. However, political realists have long argued that actors behave in self-interested ways, representing the dominant interests of individual nation-states. The brief selection of interests outlined below will suggest that vested interests are indeed prominent but that, in a globalized world, the nation-state is not the only source of these. Institutional cultures can, arguably, transcend national interests.

The Association of Small Island States (AOSIS), a group of 36 low-lying (and highly vulnerable) island states has (unsuccessfully) lobbied for a commitment to a 20 per cent reduction in emissions by 2005. For obvious reasons, these islands fear sea-level rise and favour a more cautious approach than many less vulnerable nations. Many small islands also rely on coral reefs for fishing and tourism industries and these reefs are proving to be highly sensitive to rising ocean temperatures.

'G77' is a UN-based consortium of Third World countries and a counter to the core nations' G7 (now G8) club. From 1995, many G77 nations, including China and India, have supported the AOSIS mandate: they have come around to thinking that those countries with the greatest responsibility for climate change should take strong action to ensure that others don't suffer the consequences. G77 countries have also pursued further dimensions of a **North–South agenda**, bringing issues such as trade, aid and debt into the debate. Key principles have included their right to industrial development and, as a result, to *increase* their emissions of greenhouse gases. Pursuing Brundtland-style commitments to equity, some suggest that the UNFCCC objective should include a commitment to per capita emissions convergence, i.e. a future in which per capita emissions of greenhouse gases are roughly the same for all countries of the world. This inspired objective would replace the assumption of continued disparity that is central to most trajectories of the future (including the IPCC's). Finally, Third World environmentalists such as Vandana Shiva (1993) express concern about 'green imperialism', attempts by core countries to employ the smokescreen of common interest to assert new forms of control over environments in the periphery.

The United States' political structure has proved an obstacle to decisive presidential action because Congress is very sensitive to national and state-level industrial lobbies (Paterson, 1996). By late 2000, Congress had yet to ratify the Kyoto Protocol and, as action continues to be delayed, the commitment to 7 per cent reductions over 1990 levels becomes ever more challenging. Between 1990 and 1998, United States' emissions actually grew by 11.6 per cent and the Kyoto Protocol now represents something like a 19 per cent reduction by 2008–12 (Table 8.4). The United States has allied with major oil-producing countries, attempting to slow down the UNFCCC process and water down any commitments. It has lobbied strongly on three

Table 8.4 Greenhouse gas emissions in the 1990s (million metric tonnes CO_2 equivalence)

	1990	1991	1992	1993	1994	1995	1996	1997	1998	Change (%)
US	1,632	1,620	1,645	1,675	1,713	1,733	1,790	1,813	1,820	+11.6
UK	616	620	605	590	584	575	594	568	nd	−7.8
Australia	385	387	388	391	395	408	419	nd	nd	+8.8

Note: UK reductions have largely been met through the switch from coal to gas-fuelled power stations, a move unrelated to greenhouse gas policy and a key factor in bringing the UK 'on board' during the 1995 Berlin and 1997 Kyoto negotiations.
Source: Data from National Greenhouse Gas Inventories.

issues. Firstly, it has unsuccessfully lobbied for target commitments from developing countries. Secondly, it has argued that the United States should be able to invest in net emission reductions where it will get the best results for its dollars. For example, it wants to be able to achieve a substantial amount of its reductions by planting trees or building cleaner power stations in developing countries, as an alternative to more politically and economically expensive reductions at home. Such overseas activities provide 'carbon credits' that can be balanced against domestic emissions. Thirdly, it has successfully lobbied for an international quota and carbon-trading regime. This will mean that, where one country emits less than its assigned quota of greenhouse gases, the United States will be able to buy up its excess capacity, and will do this where it proves cheaper than reducing domestic emissions. The concept of carbon credits and carbon-trading are largely accepted, but the degree to which the United States should be able to rely on these, using its financial might and trading skills to avoid domestic reforms, is contested by European and G77 nations. It is this issue that led to the collapse of negotiations in The Hague in November 2000.

Well-funded and highly organized industrial lobbies such as the Global Climate Coalition (GCC) have heavily influenced United States politicians. This body represents major industries such as Exxon, Shell and General Motors, and funds scientists who are willing to challenge the IPCC consensus on global warming. Further, it has gained excellent access to the White House and ensures it is represented at all UNFCCC negotiations (Brown, 1996). The Cato Institute, a right-wing think tank that is also financed by the likes of Exxon and General Motors, published views on the Internet during the 1997 Kyoto negotiations that included assertions that global warming would, on balance, be beneficial to the United States (see Taylor, 1997). The GCC has since declined in influence and consensus on climate change is finally spreading within the United States.

Compared to the pro-carbon industrial lobby, it is probably fair to say that environmental groups have been relatively ineffective, partly because their message has been less welcome to politicians. Large international groups such as Friends of the Earth and the World Wide Fund for Nature (WWF) have been somewhat incorporated into the consensus, not wishing to be accused of

Box 8.2   No regrets

Decisions about global warming involve very high stakes. Successful international action to save the ozone layer, through the Montreal Protocol, involved the restructuring of a five billion dollar a year industry in CFCs (Brown, 1996). Phasing out, or even substantially reducing, greenhouse gas emissions is in a different financial ball park. The emission of carbon dioxide, nitrous oxide and methane is fundamental to modern industrial and agricultural production. While politicians publicly sign up to the 'precautionary principle', what they are *most* cautious about is taking expensive action against global warming that will (a) weaken their economic competitiveness relative to rival nations and (b), worse still, turn out to have been unnecessary. Nagging scientific uncertainty remains very influential and political leaders still harbour thoughts that future science will reveal that the greenhouse effect has been exaggerated. Frustrated by this form of caution, those seeking immediate responses are advocating actions that will be beneficial *regardless of the 'truth' about climate change*. For example, action to reduce road traffic, such as support for public transport, can be considered as a 'no regrets' action. It can lessen road congestion, decrease road-building budgets, protect local ecosystems, reduce health-threatening air pollution, reduce fuel imports and, as a bonus, reduce CO_2 and N_2O emissions. A recent report by the WWF entitled 'America's Global Warming Solutions' details how the United States can substantially curb greenhouse gas emissions while actually boosting its economy. Employing financial incentives, regulation and market interventions, it is claimed that the Kyoto commitment could be doubled to 14 per cent, while also saving $43 billion per year, creating 870,000 new jobs by 2010 and bringing major improvements to air quality (WWF, 1999). The potential for 'no regrets' policies has been boosted by recent data suggesting that economic growth is becoming less reliant on fossil fuel combustion. In 1998 the world economy grew by 2.5 per cent, while CO_2 emission from burning coal, oil and gas fell by 0.5 per cent. China's economy grew by 7.2 per cent, while emissions fell by 3.7 per cent (Environmental News Service, 1999).

'scare stories' and not really offering a challenge to modernist, technocentric paradigms. Even Greenpeace has found it difficult to hit the headlines because the nature of the issue offers little opportunity for media-friendly direct actions. However, groups such as the WWF have been influential in promoting cost-effective precautionary actions that will involve 'no regrets' (Box 8.2). It has largely been left to smaller groups, and especially networks of Third World activists, to employ issues of atmospheric pollution as part of a challenge to paradigms of technocentrism, neoliberalism and globalization.

Brown (1996) draws on his experience as a journalist to suggest that political leaders become strongly influenced by both peer pressure and media pressure linked to electorates. He suggests that such pressure becomes intense when it comes to critical moments on the world political stage: at the eleventh hour, even those wishing to delay action are reluctant to stand out as the ones who refuse to sign up. The public exercises an influence that is connected to the media. Research shows that there is international awareness and concern about global warming. However, understanding of the issue is often vague and, for example, many mistakenly believe stratospheric ozone depletion to be part of the same problem. Global warming is also a psychologically distant concern that does not enter directly into people's lives, except perhaps where a hurricane is (mis)reported as being caused by climate change. As a result, global warming is a concern for a majority of people but tends to come low down on their list of environmental and social priorities (see Bord, Fisher and O'Connor, 1998). Air pollution, leading to respiratory problems, tends to be of much greater concern. This is understandable because even fairly low and officially 'safe' levels of air pollution have been very strongly linked to the aggravation of conditions such as asthma, the most common chronic childhood disease in industrial countries (Clark *et al.*, 1999).

8.5 Further reading

The opening discussion of economics can be followed up with an investigation of environmental economics. There are useful chapters in Owen and Unwin's (1997) *Environmental Management* and O'Riordan's (1996) *EcoTaxation*.

For two highly contrasting views on the 'limits to growth' argument, read the following with a critical eye: Julian Simon's (1998) *The Ultimate Resource 2* provides neoclassical optimism, while P. and A. Ehrlich's (1998) *Betrayal of Science* provides the antithesis.

On the politics of global warming, try Jäger and O'Riordan's (1996) edited volume *Politics of Climate Change: A European Perspective* (Routledge, 1996) and, for a Southern perspective, Agarwal, Narain and Sharma's (1999) *Green Politics*. (The latter can be ordered from the Website below.)

8.5.1 Websites

IPCC documents can be downloaded from: http://ipcc-ddc.cru.uea.ac.uk
IPCC data from: http://www.ipcc.ch/
UK data from: http://www.detr.gov.uk/
For Southern perspectives, consult Centre for Science and Environment: http://www.cseindia.org

Chapter 9

Postmodern environments

9.1 Introduction

In Chapter 6 we introduced the idea that the capitalist world economy experiences waves of growth and decline and that each new growth period is driven by economic and social restructuring. Chapter 7 covered the last 'short wave' in the world economy, which entered decline in the 1970s. The 1980s witnessed a new wave of growth that involved such dramatic economic and social restructuring that it has led to claims of a new, **postmodern** historical era. It is this latest restructuring of the social and biophysical world that forms the focus of this chapter. In summarizing the key transformations since the 1980s, we concentrate on the historical changes in the productive use of nature and the political and cultural responses to this momentous remapping of spatial relations. The chapter finishes on a philosophical note, considering the implications of postmodern theories of knowledge for our approach to learning about environments.

9.2 Economic globalization

Forces and relations of production have been stretched over space during the modern era. Migration, imperialism, colonialism, postwar economic expansion and the international division of labour have all contributed to an internationalization of economies, driven by the expansionist and integrationist dynamics of capitalism. Correspondingly, we have seen how environments have become internationalized, not only in terms of the chains of commodity transactions that cross national borders but also in terms of the scales of environmental hazards and the corresponding political and civilian responses. Since the 1980s internationalization has been described as **globalization**. While internationalization was largely a movement of capital, politics, people and goods across borders, globalization involves the functional integration of global society (Kelly, 1999). On the one hand, globalization disperses and fragments production; on the other, there is increasing centralization of control over global activities and the evolution of a yet more highly organized sociobiophysical system.

By the mid 1970s, the **Fordist** phase of economic production, based upon the nation-state as the unit of economic and political organization, appeared to have exhausted its potential for economic growth. The returns on capital investment were diminishing and this demanded new methods for transforming nature that, in combination, enabled new forms of capital accumulation. Firstly, new types of production were facilitated by information technologies, labour-shedding robotics and biotechnology. Biotechnology was applied to the remaking of productive environments such as agriculture and also to the creation and commodification of new organisms through genetic engineering. Secondly, large corporations fought off declining profits by further internationalizing production, searching out cheap raw materials and new sinks for depositing wastes, and establishing new pools of labour, new markets and new strategic alliances. Thirdly, such companies adopted new forms of internal organization, developing specialized and flexible units of production, integrated within networked organizations.

Post-Fordist consumption patterns are based upon consumer demand for increasingly differentiated products that are more rapidly replaced by new ones: for example, car models and trimmings change more frequently, with 'special' or 'limited' editions catering for consumers' desire for difference. Post-Fordist consumption has required post-Fordist production methods: more flexible production based on small production runs, smaller production units and more adaptable production spaces and workforces. Such consumption patterns are also linked with the so-called 'throw-away society' because goods become more quickly outdated and replaced, while presentation, often in the shape of extravagant packaging, becomes as important as content. Such intensified consumerism counteracts cleaner production technologies by increasing the demand on natural resources. There is also a trend towards new ways of selling nature which appeal to consumers' increasing reflection on environmental issues. Ecotourism, for example, is the most rapidly expanding sector of the tourism industry.

These new forms of post-Fordist capital accumulation are known as **flexible accumulation**. An important feature of the shift to flexibility has been the freeing up or 'deregulation' of capital, releasing it from the spatial constraints of the nation-state and allowing it to be moved more freely across borders. Financial deregulation is exemplified by:

- a trillion dollars worth of currency traded each day;
- financial flows that are 50 times greater than transactions involving goods and services;
- in 1975, international security transactions in core economies amounted to less that 5 per cent of Gross Domestic Product; in 1995 they amounted to 1,000 per cent of GDP in the UK (David, 1997).

9.3 Political restructuring

Economic, political and social restructuring have stimulated the growth of an area of academic study known as **regulation theory**. In essence, regulation is the creation of rules and their enforcement through some form of institution.

Generally, the institutions have been public (state) bodies, such as the UK's Environment Agency, formed in 1996. Clearly, the effectiveness of regulation is vital to environmental management: we have seen that the market is not a successful regulator of resource extraction, waste management or the distribution of benefits, and therefore institutionalized regulation is essential in many areas.

In the 1980s and 1990s there was considerable attention paid to deregulation: of the economy, of capital, of public transport and so on. To some extent, this shift away from a 'command and control' state was part of the political right's neoliberal political agenda, as exemplified by Margaret Thatcher's project of dismantling the 'nanny state'. Simultaneous to this state-level programme of deregulation, there has been a growth in supranational regulatory institutions, such as the European Union (EU). Viewing these processes in tandem, some commentators have pointed to the 'hollowing out' of the nation-state, witnessed by a loss of a domestic regulatory function. However, during this period, the UK has itself seen the creation of more central regulatory institutions than ever before (Baldwin, Scott and Hood, 1998). Many of these are quangos (quasi non-governmental organizations) such as the new utilities regulators. In contrast to a political programme of deregulation, the UK state has actually increased its regulatory functions. Despite this, however, most regulation theorists are concerned about a **regulation crisis** – a reference both to the effectiveness and the accountability of the new mix of national and supranational regulators. Conventionally, regulation is intended to serve the public interest through, for example, environment, planning, labour, health and safety, and food safety regulators. However, there is growing concern that quangos and supranational institutions have no electoral mandate and lack accountability. Partly as a result, it is contested that they often fail to represent the public interest and instead serve powerful corporate interests. We examine such concerns, particularly in relation to supranational regulators, and find a loss of trust in current modes of regulation. While many of these institutions have incorporated the neoliberal politics of the new right, a growing number of the public are demanding a new basis for regulation, rooted in sustainable development.

Supranational institutional governance has gone hand in hand with a revitalized neoliberal political ideology. Neoliberals consider that the free flow of capital, goods and services across the globe will bring sustained economic growth and associated social progress. In fact, many neoliberals do not just portray this as a desirable future, but as an inevitable one (Kelly, 1999). The political ploy of pretending that the ideological struggle for neoliberalism is already won is both commonplace and powerful. It enables global institutions such as the World Trade Organization (WTO), the World Bank and the IMF to pursue neoliberal policy uncontested: if there are no alternatives, there is no need for debate and no need for decision-making to enter into the arena of democracy. In spite of this magnificent global confidence trick there is an increasingly powerful lobby for regulating the power of capital and managing the flows of investment and trade in ways that protect both people and planet. These are some of the concerns about neoliberalism and globalization:

- Mobilization of capital, together with globalized competition, leads to investment in nations with the most lax environmental standards. In the 1990s, for example, Shell was criticized for maintaining lower environmental standards in Nigeria than in its operations in core countries. Standards can be forced down in order to attract or keep investment, involving a loss of local capacity to manage environments.
- Mobilization of capital leads to investment in areas with the lowest wages, least organized workforces and worst health and safety regulations. Of the £100 paid for a Nike ski jacket in the UK, it was found that just 51p was earned by the Bangladeshi women who made it (Robins, 1999).
- Globalization perpetuates and deepens gross inequalities of wealth. During the 1990s, the poorest 20 per cent in the world saw their share of global income decline from 2.3 per cent to 1.4 per cent (Giddens, 1999). In 1960, the richest 20 per cent earned 30 times more than poorest 20 per cent; in 1997, they earned 74 times more (Robins, 1999).
- Unrestricted global commerce benefits the already affluent. Between 1986 and 1996, Ghana increased its cocoa exports by 80 per cent but its income from this trade grew by just 2 per cent (Robins, 1999).
- Globalization leads to the homogenization of cultural landscapes through the Westernization and 'coca-colonization' of places (Peet, 1989).

Used as a political tool to prevent responsible regulation of markets and capital, the neoliberal globalization discourse is fundamentally undemocratic and runs contrary to the needs of sustainable development. While capital has become increasingly global, so too have the political and corporate decision-making bodies that undertake global organizational functions. Giddens (1984) describes a process of **time–space distanciation** in which the decisions that affect people's lives become increasingly distanced from them. Time–space distanciation is a feature of globalization that is facilitated by the expansion and speeding up of communications. Important decisions are increasingly 'pulled away' from local communities and centralized in state governments, global institutions and the boardrooms of multinational corporations. For example, the Kurdish people living in the vicinity of the proposed Ilisu dam in southeast Turkey had their livelihoods hanging by a thread at the time of writing. The decision whether to flood their lands will not be taken locally; it will depend on the Turkish government, the leverage that opponents in Syria and Iraq can muster and, most distanced of all, whether the British government goes ahead with its plan to provide the construction company Balfour Beatty with £200 million of export credit guarantees. Globalization can thus be fundamentally disempowering, commodifying environmental resources within physically and psychologically distant chains of transactions and short-circuiting local knowledge and management.

The WTO exemplifies a powerful global institution that has been given sanction to wield the 'no alternatives' discourse to promote neoliberalism. As an international institution, the WTO lacks the democratic accountability of a state and yet is a binding authority that can force compliance with its trade

Figure 9.1 Protest against the WTO and global capitalism, London, November 1999
Source: Adrian Martin.

liberalization agenda. The WTO regularly rejects calls for market regulations that seek to uphold social and environmental rights and is frequently charged with repressing local people's rights to protect their local economies and environments. The WTO has become a target of environmental and social campaigning, as witnessed by worldwide protests during its November 1999 meeting in Seattle. Prior to the meeting, over 600 environmental and human rights organizations signed a petition seeking to limit the WTO's powers.

9.3.1 Agency and resistance

Globalization, with its new configurations of time and space, has involved the pulling away of power from the local, disempowering individuals and communities and constraining their ability to manage local environments. Thankfully, there is also a more positive process to describe. The communications revolution has contributed to **time–space compression** (Harvey, 1989a), a process which connects the world in ways that reduce the time it takes for money, information and people to flow from one place to another. Such increasing connectivity (the 'shrinking world') can help to connect local communities with the global, enabling them access to information, to the media and to like-minded individuals and groups.

In a globalized world, access to information, and the networking of responses to information, is a powerful tool, as demonstrated by the opposition to the proposed Multilateral Agreement on Investment (MAI). This was an attempt by the United States and Europe to liberalize investment further, seeking to augment global 'regulation' of free trade (the WTO) with global regulation of free investment. The MAI would have provided multinational corporations with new powers over the nation-state, limiting the sovereign right to regulate inward investment just as the right to regulate imported goods and services has been degraded by the WTO. For example, the MAI would have made it difficult for nation-states to impose particular, nation-specific, environmental regulations on investors seeking to locate production activities within their borders. In fact, it introduced mechanisms whereby companies could sue governments for introducing stringent environmental standards. While such a move was again wrapped up in the discourse of 'no alternatives', the resultant backlash from environmental, labour, women's and human rights groups proved that globalization could be resisted. A global network, loosely and flexibly connected by time–space compressing technologies, circulated information about the dangers of the MAI and then successfully opposed this threat to environmental democracy. MAI negotiations collapsed in 1998 following the withdrawal of France. Thus, while the global can be an arena in which local communities feel powerless, it can also be the arena in which social and environmental progress is made and civil society can flourish. Globalization provides the most appropriate scale for analysing environmental issues and, through communicative integration and counter-politics, unleashes the global potential for progressive social movements: globalization can be both constraining and enabling.

9.4 Risk society

For Beck (1992), such counter-politics are a necessary response to unacceptable environmental risks and a loss of trust in existing regulatory mechanisms. The environmental hazards unleashed by modern technology, combined with increasingly distant and unaccountable decision-making, literally forces people into resistance. Despite evidence of increasing longevity and personal security for those living in the richer nations, the environmental risks currently being released are taking on enormous cultural significance, arguably more than at any previous juncture in history. This is described as the transformation of industrial society into **risk society**, a state that is characterized by a new breed of risks. Old risks include diseases, earthquakes, hurricanes, floods and droughts. New risks include nuclear fission, atmospheric pollution and the creation of new chemical and biological forms.

New risks differ from old risks in the following respects:

- New risks are knowingly and wilfully imposed upon some people by other people. For example, the decision to allow field trials of genetically modified foodcrops involves dominant economic interests imposing experimental risk onto often unwilling participants.
- New risks have less spatial limitations. The release of persistent pollutants such as polychlorinated biphenyls (PCBs) leads to transnational and even global risk.

- New risks have less temporal limitations. Issues such as radioactive pollution and land-degradation create hazards for future as well as current generations.
- New risks have less social limitations. Old risks, including diseases such as malaria and natural disasters such as floods, affect the poor far more than the wealthy. Certain new risks, such as nuclear meltdown, are rather more democratic.
- Some new risks are potentially apocalyptic. Global warming, nuclear war and even newly manufactured diseases all raise the spectre of humanity's nemesis.
- New risks often arise from attempts to deal with past crises. For example, concerns about fossil fuel scarcity and carbon dioxide emissions led to subsidies for the nuclear industry that in turn led to the increased risk of radioactive pollution.
- The causes of new risks are often distanced from their impacts, making it harder to attribute blame. The origins of some disasters, such as the AIDS pandemic, remain uncertain.
- The ability to attribute, calculate, insure against and compensate for risks has diminished. New risks cannot be managed through the same instruments and institutions as old risks.

Risk society is characterized by the **organized irresponsibility** of governments, corporations and scientific communities (Goldblatt, 1996). This is the intentional and unintentional concealment of the real causes and consequences of environmental hazards. When faced with an alleged risk, the primary act of irresponsibility is to employ state or corporate scientists to deny the possibility of a link between the alleged cause and effect. In the past, 'scientists' have been employed to claim: no proven link between BSE and CJD; no proof that nicotine is addictive; no proof that sulphur dioxide contributes to acid precipitation; no proof that genetically modified crops cannot be contained within selected field sites. For accidents and catastrophes, a favoured tactic is to blame human error. This enables the scientist to maintain that the technology is safe and thus to deny any genuine systemic risks.

Organized irresponsibility has contributed to the **fragmentation of science** because official science becomes less trusted and more frequently challenged. Environmental conflicts can see expert pitted against expert in a confusing battle for legitimization: government-sponsored scientists tell the public that nuclear power and GM foods are safe, Greenpeace scientists tell them otherwise. Such exchanges can contribute to the **delegitimization of science**, a process that undermines a central feature of modernity. Delegitimization of science breaks the link between science and progress (Lash and Urry, 1994; Barry, 1999), eroding the discourse of Enlightenment. It also creates the cultural space in which alternative sources of knowledge can flourish. Box 9.1 examines organized irresponsibility and the delegitimization of science in relation to the UK's Sellafield nuclear facility.

For social theorists such as Ulrich Beck, Anthony Giddens and Jurgen Habermas, the response to the environmental crisis must embrace forms of

Box 9.1   Sellafield, Cumbria

Sellafield represents the modernist, technological dream gone wrong. Since 1952, a military reactor has produced weapons-grade plutonium; in 1956 the world's first nuclear power station opened (it is still operating); since 1994, the Thermal Oxide Reprocessing Plant (THORP) has imported spent fuel from Europe and Japan, creating a multibillion pound industry and thousands of jobs.

On 10 October 1957 fire broke out in the fuel piles and a cloud of fission products entered the atmosphere. A change of name (from Windscale) did little to change the image. Since 1952, Sellafield has dumped radioactive waste into the Irish Sea, leading to one of the most radioactive marine environments in the world. A small proportion of this radioactive waste finds its way back to land via sea spray and it has been found that the closer you live to Sellafield, the greater the concentration of plutonium likely to be found in your body (O'Donnell *et al.*, 1997). Emissions peaked in the late 1970s, then entered a long period of decline but are thought to have started rising again since operations at THORP began in the mid 1990s. Reprocessing creates huge volumes of high-level radioactive waste.

British Nuclear Fuel's handling of perceived risks shows signs of 'organized irresponsibility'. A good example of this is the handling of scares over 'leukaemia clusters'. In the 1970s locals observed an apparently high level of childhood leukaemia in the vicinity (Wynne, 1996). BNFL went into denial, arguing that Sellafield's emissions were small compared with natural 'background' radiation and posed no risk. When environmental groups attempted to investigate, the government denied them access to medical data. Denial, and attempts to suppress independent research, continued until 1983 when a television documentary aired local concerns. As a result, the government convened a select committee whose report found that there was a cluster of excess childhood leukaemia in the vicinity of Sellafield but that this could not be scientifically linked to the nuclear plant. With disregard for the precautionary principle, Sellafield remained innocent until proven guilty, until local people could scientifically prove a link. In March 1996 the government concluded its research into child cancer clusters: there was officially no link and Sellafield had no case to answer. Across the sea, the Irish government reported increased levels of cancer along its east coast and became similarly frustrated by demands that it must bear the burden of proof. The Irish people are faced with an even more inaccessible and distant source of imposed risk, one from which they don't even derive economic benefits.

Sellafield's future may now be measured. Not because environmental democracy has won out or because a new morality has been brought to bear on economic policy and corporate activity, but because BNFL have made some dismal economic decisions and have lost the trust of client governments. To start with, the THORP facility has never come close to its target of reprocessing 7,000 tonnes of spent fuel in its first ten years. The waste is

(continued)

(continued)

being piled up in vast cooling ponds and there is no market for the plutonium end product (Brown, 2000a, 2000b; Wintour and Wainwright, 2000). The Mixed Oxide (MOX) plant looks like being even more of a white elephant than THORP. It was completed in late 1997 at a cost of £300 million of public money. However, it has not been granted a licence to commence operations because there is insufficient evidence of market demand.

In September 1999 falsified quality-control documents were discovered. BNFL insisted that a recent shipment of fuel pellets to Japan was safe, but a subsequent investigation by the Nuclear Installations Inspectorate found this to be untrue. BNFL maintained the show of denial for three more months before the Japanese discovered more falsified quality documents, relating specifically to their shipment. In February 2000 it emerged that Germany had also received suspect fuel pellets from Sellafield and, worse still, these had been in a reactor since 1996. The reactor had to be shut down, at great cost. BNFL sought to allay fears about systemic risk by blaming human error: the chief executive and five quality-control workers were fired. But they were not winning back friends: the Swedes were next to pull out, followed by the Swiss. Even British Energy announced a desire to shift from reprocessing to storage of waste.

---

environmental democracy, defined by Mason (1999) as participatory and ecologically rational forms of collective decision-making. As a first requirement, this must involve democratic processes breaching areas of decision-making that have previously been depoliticized: scientific research and development, economic enterprise and administrative bureaucracies (Mason, 1999; Goldblatt, 1996). New social movements (NSMs) are partly a response to the truncation of democracy. For example, the alliance fighting the WTO are united in the ambition to bring the issue of trade back into the arena of open, democratic governance (Robins, 1999). In addition to the *extensification* of democracy and citizenship, to incorporate issues of social and environmental justice, there is also a need to *intensify* them through the development of participatory, in addition to representative, forms of democracy (see Chapter 12).

9.4.1 A new morality?

For Giddens, Beck and Habermas the emergence of NSMs, and especially the environmental movement, signals the emergence of a new morality. The space for such a new morality has been opened up by a loss of faith in the modern project. NSMs are seen as evidence that this moral void is being filled by new commitments to universal norms based on social and environmental justice. Individuals are seen increasingly to monitor the wider implications of their personal activities and to do so in ways that bring morality to bear on economic and political decision-making. This is what is termed **reflexive modernization**.

This should partly be thought of as a 'reflex' reaction to new environmental dangers, i.e. resistance and protest as a kind of unthinking self-defence mechanism. More importantly, perhaps, it should be thought of in terms of a conscious 'reflection' upon personal day-to-day actions, within the framework of new moral commitments. The aggregate of these commitments is the emergence of powerful social movements, stretched across space and structured by informal, self-organizing and evolving networks.

Theories of reflexive modernization are problematic for many postmodern theorists. They see the identification of a new universal morality and global social movements to be unfounded. Postmodernists tend to focus on processes of fragmentation rather than processes of integration (such as the formation of broad-based movements) and are highly sceptical of theories that claim to explain social formations. Whichever view of NSMs you subscribe to will be dependent on whether you think they amount to an integrated social dynamic and whether you believe that social change is in fact characterized by such broad-based movements.

9.5 Postmodern culture

We have seen that postmodern economic restructuring involves the stretching of money, power and commodity chains across space, releasing them from their old physical and political constraints. While this can lead to the integration and homogenization of cultures and landscapes, the opposite is also the case: globalization and postmodernity are consistent with the enhanced differentiation of people and environments. The fact that Coca-Cola, Levis, Star Wars, Microsoft software and Puff Daddy can be consumed globally does not mean that cultural landscapes necessarily become similar. We wouldn't, for example, expect the availability of tens of thousands of music CDs to result in everyone having a similar collection. Indeed the wide availability of cultural artefacts can result in new and exciting combinations, cross-cultural hybridization (music, gardens, architecture) and places that contain ever more complex and unique cultural mosaics. Furthermore, individuals and places are not empty vessels waiting to be given meaning and identity through the consumption of goods and services: a location does not necessarily become 'America-like' just because people there consume Big Macs and 'Baywatch'.

To believe that places are 'different' because they are somehow protected from outside cultural influences is somewhat of a myth. Even apparently 'authentic' places, such as a remote village advertised as 'the real Spain' have been consistently influenced by geohistorical encounters with the wider world. The physical–cultural landscape of such a village may have encountered some or all of the following: Roman conquest and settlement; Moorish conquest and settlement; trade with South America; international conventions of architecture; agricultural landscapes influenced by multinational chemical companies and the Common Agricultural Policy; a road built with European Union money; agricultural irrigation systems inherited from North African designs; vineyards with Cabernet Sauvignon grapes, a graveyard with the bodies of Irish or American workers who died during the Civil War. Local places are inextricably linked (historically and geographically) with 'outside' places and are always influenced by other cultures. So difference is not just a product of being bounded off and isolated from the world. Local difference is actually created and perpetuated by the ways in which the global is incorporated into the local and the local is incorporated into the global. As Crang (1999) argues, local communities confer their own meaning onto imported cultural artefacts and, in doing so, reconstruct them in localized ways. Globalization speeds up and extends processes of cultural hybridization, arguably leading to more fragmented and differentiated landscapes in which local difference is not just tolerated but celebrated.

In a similar way, globalization contributes to the fragmentation of individual identities, which become less constrained by the circumstances of birth (class, ethnicity, gender, location, and so on), and which can be reconstructed through the customized consumption of commodities. Globalization has enabled this fragmentation, unleashing the ability of people to construct and define their own identities through the selective consumption of a vast array of global cultural artefacts. Just think through what you have eaten today, what clothes/fabrics you are wearing, what music you listen to, television you view, Internet sites you access, and so on. Thinking backwards a little, Beck and colleagues were suggesting that forms of integration and organization (such as an environmental movement) actually emerge amid this fragmentation: order emerges out of chaos. We will later see that they might find some support for this from the unlikely source of science.

Postmodernism is sometimes used to describe a new aesthetic involving a reorientation of what we find beautiful in environments. Most notably perhaps, the shift from modern to postmodern architecture has involved a reassessment of environmental aesthetics. Modern architecture was influenced by the beauty of function and geometrical science (Powell, 1998). Le Corbusier reduced design to bare essentials, emphasizing clean lines and pure, simple geometrical forms. But postmodernity has put paid to this vision: landscapes of concrete and glass fail to differentiate the local, while geometrical shapes idolize defunct scientific certainties about nature. Postmodern architecture combines international styles and conventions with a sensitivity to the cultural heritage of place, creating and emphasizing difference through local motifs and by creating hybridized landscapes that are tolerant of a wide variety of people.

Cosgrove (1990, p. 354) cites the English art critic Peter Fuller to explain the significance of postmodernism to our aesthetic appreciation of nature:

Fuller, in an essay titled 'Neo-Romanticism' notes a significant shift in the artistic predilections over the past decade in England and elsewhere: 'The return to landscape has become something of a stampede. Ten years ago, no self-respecting art student . . . would have touched a box of watercolours or have gone near lakes, valleys, rolling fields and small Gothic churches. Today, the hills are alive with the sight of *plein air* painters once again' (Fuller, 1985, p. 83). Modernism in art, he argues, drove a wedge between the pursuit of art and the study of the natural world. If, as the Modernist philosophy argued, nature lacked mind, and even the divinity that Romantics like John Ruskin sought to read in nature did not exist, then nature was unworthy of artistic attention or consideration – it was quite literally unaesthetic.

9.6 Has science gone postmodern?

In reviewing the evidence for postmodern economic, political and cultural processes, 'science' has been characterized as a bastion of modernity. And yet many self-avowedly postmodern environmentalists draw heavily on scientific progress. Here then, we consider the possibilities for a postmodern science.

Modernist, reductionist scientific methods have proved remarkably successful at describing nature and as a basis for material progress. Reducing the study of nature to the analysis of component parts has enabled great advances in the fields of medicine, agronomy, metallurgy and engineering. However, while this form of scientific enquiry may have served well over the centuries, its limitations are now becoming apparent. In the 1920s Heisenberg's Principle of Indeterminacy demonstrated that, when investigating sub-atomic systems, you cannot measure one part of the system without rendering another part immeasurable. For example, you cannot measure the position of a particle without rendering its momentum immeasurable: if the researcher wants an electron to have particle-like properties, it will; if the observer wants it to have wave-like properties, it will (Matthews, 1991; Capra, 1982). In opposition to the doctrine of positivism, it appeared that what was observed as reality was in fact contingent upon the observing scientist.

Such concerns about subjectivity in science have pretty much been shaken off and positivism has survived intact (Gandy, 1996). However, since the 1950s the limitations of modernist science have been further exposed by observations of apparently random, chaotic behaviour in nature. By definition, that which is random is not amenable to prediction and certainty and therefore sets limits to scientists' ability to understand and describe nature. However, 'chaos' should no longer be interpreted as nature acting in arbitrary and disorderly ways. The popular chaos exemplar, a storm in Rio resulting from a butterfly flapping its wings in Beijing, provides an awesome vision of nature. However, this example portrays the least fantastic aspect of chaotic behaviour. What really amazes the new scientist is that, in nature, order frequently emerges out of chaos. Butterfly wings are flapping all over the world, all of the time, but, significantly, this does not lead to a completely anarchic and disorganized global climate. Despite millions of butterfly wings, there are discernible patterns of organization in the global climatic system (even if these patterns are too complex to be fully understood). When we look at the small scale, at the butterfly, we see random behaviour; when we look at the whole picture, at global climate, we see the emergence of ordered patterns of behaviour. Scientists describe this as the emergence of **order out of chaos**.

Examples of complex forms of system organization abound in nature: weather systems, patterns in sand dunes, the wave behind a boulder in a stream (further examples were cited in Chapter 2). The most remarkable patterns of organization can be found in living organisms. For example, you can't understand the property of 'life' through a reductionist analysis of a cell's biochemical constituents; ditto for the property of 'consciousness' in human brains. These are examples of what are called 'emergent properties', properties that cannot be observed through analysis of parts because they are a property of the organization of the whole and are therefore only amenable to holistic observation. James Lovelock's 'Gaia theory' is perhaps the ultimate in holism: it seeks to identify the emergent properties of planet Earth and is sometimes characterized as postmodern science (Box 9.2).

Box 9.2   Gaia

The 'new science' of complexity focuses on the dynamic behaviour of whole systems, recognizing how component parts interact with each other to produce overarching patterns of behaviour, or emergent properties. While most scientists have restricted themselves to studying complexity in systems with few variables and simple rules, the British climate scientist James Lovelock, together with the biologist Lynn Margulis, have attempted to study the emergent properties of the Earth itself: an undertaking of breathtaking ambition! Lovelock began to envision the Earth as a self-organizing system while working for NASA's space programme and contemplating the best way to detect life on Mars. Lovelock realized that life was best indicated by the existence of high amounts of unstable gases, such as oxygen and methane, in a planet's atmosphere. Pursuing this line of thought, he puzzled over how the temperature of the Earth had remained relatively stable despite the atmosphere being highly unstable and despite the fact that the temperature of the Sun was known to have increased by 25 per cent during the four billion years of life on Earth.

> It was at that moment that I glimpsed Gaia. An awesome thought came to me. The Earth's atmosphere was an extraordinary and unstable mixture of gases, yet I knew that it was constant in composition over quite long periods of time. Could it be that life on Earth not only made the atmosphere, but also regulated it – keeping it at a constant composition, and at a level favourable for organisms? (cited in Capra, 1997, p. 102).

This glimpse of Gaia led to a theory of the Earth as a complex, self-regulating system in which all the component parts (animals, plants, rocks, soils, air and so on) interact to produce emergent patterns of organization that, though far from equilibrium, are relatively stable. His neighbour and friend in Cornwall, the novelist William Golding, suggested the name 'Gaia', after an ancient Greek earth goddess. Looking at the whole picture (at Gaia), Lovelock and colleagues were able to identify emergent properties of the system: feedback mechanisms that led to the regulation of climate, moderation of ocean

(continued)

(continued)

salt levels, regulation of atmospheric oxygen and carbon levels, and high levels of biodiversity. The regulation of Gaia's climate is partly analogous to thermoregulation within the human body. Body temperature is far from equilibrium: it is vulnerable to internal and external changes and, if it falls or rises by more than a few centigrade, the result is death, followed by further temperature change. And yet it remains remarkably stable because changes lead to negative feedback loops: if your blood starts to warm, cooling mechanisms are triggered: increased sweating, hairs flattening against the skin and greater blood supply to the body's periphery. Gaia's emergent properties are, to Lovelock, feedback loops that preserve certain qualities such as climate. If one variable tends towards a warming effect, e.g. rising atmospheric carbon dioxide, this will stimulate another variable to oppose this trend. For example, increased microorganism growth in the oceans, leading to greater sedimentation of carbon-bearing shells on the ocean floor and the locking of carbon in sedimentary rocks. The comparison between Gaia and a living animal is only an analogy: animals have centralized coordination systems (a brain); Gaia clearly does not. However, some of Lovelock's readers have viewed the planet as a super-organism.

Initially, Lovelock's theory was not well received by the scientific community, not least because it posed enormous challenges to orthodox evolutionary biology. Many peers misread the Gaia hypothesis, claiming that Lovelock and colleagues were claiming the Earth itself to be a sentient being (to have a brain) or, alternatively, that Gaia's systems of self-organization must have been created with a particular evolutionary direction in mind. Leading journals such as *Nature* and *Science* refused to publish Lovelock's works. However, the new science of complexity has gone a long way towards vindicating Lovelock's general approach. The investigation of emergent system behaviours, at the whole-system level, has proved fruitful, for example in identifying the interactions between atmospheric gases and microorganisms. It should also be added that Gaia theory has proved popular in the public imagination and Lovelock has become one of the best-known living scientists.

Sources: Coveney and Highfield, 1995; Capra, 1997.

The new science has an ambiguous relationship with postmodern theory. On the one hand the identification of indeterminacy and random behaviour have fuelled postmodern accounts of nature because they seem to undermine the scientific certainties of modernity. On the other hand, postmodernists who attempt to employ contemporary science as the basis for scientific relativism, or the fragmentation of science, are poorly informed. For, as we have seen, chaos is an increasingly outdated discourse of nature. Chaos is now seen as a determinant of order rather than an end in itself. In this way, the science of complexity is in fact a form of environmental realism (Byrne, 1998) because it maintains that there is a real nature that is potentially knowable through scientific enquiry. We will see below that this contradicts much postmodern theory. As Gandy (1996) concludes, it is difficult to argue that science is undergoing a postmodern turn.

9.7 Postmodernism and epistemology

Instead of providing a neat definition of the term 'postmodernism', we have described a set of characteristic processes. This is partly because the term 'postmodern' is used in a variety of ways and its meaning is itself fragmented. Nevertheless, we have alluded to **postmodern theory**, and this is suggestive of some overarching and fundamental beliefs held by those who define themselves as postmodernist thinkers. Lyotard defines postmodernism as 'incredulity towards metanarratives' (cited in Powell, 1998). By metanarratives, he means large-scale theories about society and nature: Marxism, evolution, positivism and many more from all areas of academia. By incredulity, he implies a lack of faith in both the validity and the intentions of such theories.

In place of grand, universal theories, postmodernism offers a more fragmentary understanding of the world in which there are, potentially, as many right ways of interpreting the world as there are individuals to perceive it. Thus, postmodernism replaces the desire for universal truths with a celebration of diversity. To a limited extent, we have previewed arguments for such relativism. In the last chapter, we saw how different scientists can present different versions of the 'truth' about global warming. Employing the same 'objective' data, they can arrive at different claims to truth because there is a subjective (and power-related) component to knowledge creation. Similarly, in this chapter, we have identified the fragmentation of science, a process that arises directly from its subjective component and the loss of universal status. However, our own examples only go as far as to suggest that the *interpretation* of reality is prone to such contingency, i.e. that the way that we seek to know about reality (our epistemology) can be relative to our life circumstances. Importantly, this leaves the door open for methods of synchronizing our interpretations, largely through debate, and reaching agreement on how, for example, we should understand phenomena such as global warming. Such activities are central to the development of communicative rationality (Chapter 2).

Postmodern theorists often take a much deeper step into relativism by challenging the very existence of a reality that is independent of the human mind. In other words, they possess both an epistemology *and an ontology* that are steeped in subjectivity. Thus, while we might contest that different people, from different parts of the world, might represent a tree in different ways, postmodern theorists might contest that there is no such thing as a real tree. Such a belief underpins the 'incredulity towards metanarratives'. Quite simply, if there is no single reality 'out there', there can be no grand theories that explain how it works. What we are left with is a profound relativism, the view that truth is in the mind of the beholder.

So why do postmodernists reject the existence of a real nature in favour of an entirely socially constructed one? The answer lies in the way in which we represent nature, particularly through our use of language to assign meaning. Language is a cultural construction and anything that we understand through language takes on the same status. Let's borrow the analogy of a set of traffic lights: there is no 'reality' in which a red light actually means 'stop'. Rather, we culturally assign meaning to it and thereby culturally construct a collective knowledge of it: that 'red' means 'stop' is a convention not a reality. It is possible that someone reading this book comes from a part of the world where 'red' means 'go'. Traffic lights use colours; a language such as English uses words. 'But, like colours, words mean something only through convention' (Cohan and Shires, 1988, p. 4).

It is this kind of linguistic argument, based on the lack of a genuine link between language and an independent reality, that is central to the postmodern rejection of realism. Words such as 'tree' or 'global warming' only have meaning through convention, and therefore all natural phenomena – and all environmental problems – are cultural constructions. Perhaps, for example, the words 'hole in the ozone layer' conjure up the image of a neat circular gap in some kind of sheet-like fabric. Perhaps a friend reading the same words will be thinking of a wispy void in a cloud-like substance. The problem for postmodernists, is that they are just words, just cultural constructs, and in the absence of any other means of representing nature (images, based on colours, suffer similarly) there can be no way of objectively knowing about nature. In fact, to assert your interpretation over that of others is an attempt to assert power over them and dominant interpretations go hand-in-hand with power relations: such as men's interpretations over women's and Western science over indigenous knowledge.

The relativist rejection of grand theories has been beneficial in some respects, opening up the intellectual space for the serious consideration of previously marginalized perspectives. However, our problem here – and a problem shared by many environmentalists – is that the denial of a real nature presents a highly problematic basis for environmental management (e.g. Soper, 1999). To tie this chapter up, and offer some way out of the chasm, we return to critical realism. Gandy (1996) considers that the tension between postmodern theory and environmentalism, in particular the stand-off between relativism and realism, occurs because of the lack of distinction between epistemology and ontology. We have attempted to address this above and, in doing so, have identified more precisely where critical realism departs from postmodern relativism. Critical realism seems to offer the greater prospects for environmental understanding and action because it escapes the void of ontological relativism by accepting a reality whose existence is more than mere cultural construction. There is a real, independent nature that we act upon and, indeed, that can act upon us. At the same time, critical realism avoids falling into the naïvety of an epistemology that holds scientific enquiry to be entirely objective and unproblematic. This makes dealing with environmental problems difficult, because we accept that there will be many perspectives, or discourses, and that all of these have a contribution to make. However, as stated above, it does leave the door open for forms of communication and debate that could lead to a coming together (a synergy, hybridization or just a compromise) of views. The following chapters look into ways in which such communication might be developed.

9.8 Further reading

Barry's (1999) *Environment and Social Theory* contains an accessible chapter on postmodernism. Goldblatt's (1996) *Social Theory and the Environment* is a more challenging but rewarding text. Lyon's (1999) *Postmodernity* provides a more wide-ranging understanding.

The introduction to Baldwin, Scott and Hood's (1998) *A Reader on Regulation* gives a good overview of current trends and issues in this field.

Chapter 10

Postmodern forest environments

10.1 Introduction

We have seen that what passes for knowledge about nature has become fragmented and contested in postmodern societies. Individuals construct their own views about environmental issues and these are dependent, for example, on their cultural heritage, life circumstances, media exposure and experience of environmental risks. As a result, contemporary environmental issues are frequently characterized by competing discourses. **Discourses** are simply ways of interpreting the world around us; they are the aggregate of various beliefs and experiences, as well as the linguistic devices used for their representation. Discourses contribute to individual and group constructions of environmental problems and their potential solutions. This chapter draws on two case studies to explore the notion of 'competing discourses' in postmodern environments. Firstly, we consider some of the main discourses operating in disputed North American forest lands, with a particular emphasis on British Columbia. In the previous chapter it was noted that processes of globalization can pull power away from local communities, whose discourses become marginalized by increasingly distant decision-making institutions. Conversely, it was also suggested that globalization can create the impetus to push power downwards towards communities, reinvigorating the multiple discourses of local civil society. The case study of British Columbia demonstrates such a pushing down of decision-making, showing how new systems of community involvement, based on local institutional arrangements, are a necessary step towards developing sustainability. The second case study looks at developments in Indian forest management where, since the early 1990s, new forms of community governance have been identified as the only realistic way to protect and enhance forest resources. Global institutions such as the United Nations, the World Bank and the Intergovernmental Panel on Climate Change are long-established forms of supranational governance. The 'new institutionalism' of local, community-based arrangements for environmental management are often less well established and we currently know less about what makes them function effectively. Some of the problems of such institutions are considered in relation to India's village forest committees (VFCs).

10.2 Forestry: a modern profession

Chapter 6 discussed ways in which colonial powers conquered premodern societies, drawing them into the modern relations of the emerging world system. Such incorporation led to the decline of existing systems of forest management and, crucially, the erosion of the social capital that had enabled premodern communities to regulate the use of local resources. European settlers in North America and British administrators in India were beguiled by the sheer size of the territories and the apparently limitless resources that they were colonizing. In the absence of any thoughts of scarcity, conservation was initially of little concern and legislation focused on ensuring colonial gain. In North America, the first forest policy was intended to supply the Royal Navy with timber, with the first shipment of white pines leaving New England in 1634 (Hardy, 1997). In India, the first forest policy was an 1806 proclamation claiming British sovereignty over forests and trees, again related to the need to secure supplies of timber for military and merchant ship-building. Colonial plunder, together with rising domestic demand, inevitably overwhelmed the belief in nature's plenitude. In the early 1860s the Secretary of State for India commented on the 'grave position of the accessible forests, which were threatened with total annihilation' (Saldanha, 1996, p. 1864). In Canada, the Morgan commission of 1884 called for urgent action to stop indiscriminate cutting of forests and to control forest fires (CFS, 1999).

In both India and North America, the eventual response to new perceptions of scarcity was the application of modern forestry, institutionalized through new government departments and legislative acts. The Imperial Indian Forest Service was established in 1864 and the US Forest Service in 1905. In Canada, the Forestry Branch of the Department of the Interior was founded in 1899, later to become the Canadian Forest Service. The scientific basis of Canadian forestry was supported by the Canadian Society of Forest Engineers – a title that seems to draw on the modernist metaphor of 'nature as machine'. In both North America and India a very modern profession had emerged. Its main aim was to produce sustained yields of commercially valuable timber for the sovereign state, its underlying faith was the ability to employ scientific methods to create stable, predictable and productive forests, replacing ancient stands of mixed age and species with monoculture plantations. The transition to modernity involved a simultaneous shift from diversity to homogeneity for both forests and forest relations of production. **Scientific forestry** required standardized silvicultural systems and standardized tenure arrangements. This required that the diversity of local management arrangements were replaced with centralized authority. In particular, remnants of community management, based on premodern systems of communal tenure, were replaced by state control of forests. In this way there was a direct relationship between the subordination of non-human nature and the alienation of indigenous people from local resource management.

10.3 Forestry in North America

Gifford Pinchot, first chief forester of the US Forest Service, developed an enduring vision of how Americans should manage their renewable resources. His own words reveal a personal discourse about forests that is firmly rooted in modernity and, in particular, the Enlightenment ideal of the application of science for human progress. In his 1901 work *The Fight for Conservation*, he proclaimed that 'the first duty of the human race is to control the earth it lives upon' and 'the first fact about conservation is that it stands for development' (Pinchot, 1999, pp. 187–8). The latter comment is revealing of the still dominant view that non-human nature is only valuable in its instrumental use for human material improvement. His guidelines for conservation were based on **wise use** rather than preservation.

Through his development of a professional forest service, Pinchot successfully institutionalized these beliefs and a hundred years later, the language of 'wise use', 'balanced use', 'scientific management' and 'sustainable yields' has stood the testimony of time. North American forests continue to be scientifically managed to achieve certain outcomes that are considered useful to human well-being: predominantly timber production but increasingly tourism, recreation resources, scenic value and wildlife conservation. However, many would be quick to point out that such 'wise use' has not prevented forest loss, has not protected jobs in the forestry industry and has increasingly come into conflict with preservationist discourses.

10.3.1 Challenging scientific forestry

By itself, science cannot provide an adequate basis for forest management. Science is a vital tool for the execution of management plans but it is not a sufficient basis for creating such plans in the first place. Planning and management require values and objectives, regarding what are the best things to do with our forests. Such value judgements must be socially constructed, drawing on economic, political and spiritual priorities, as well as ecological and technological possibilities. In North America important 'constructions' of values have arisen from within a number of discourses, leading to competing visions of how forests should be managed. Before describing some of these, we need to provide some background to an important strand of environmental thinking that has not yet been explored in detail. This is a lineage of environmental philosophy that begins with the English Romantics and then cross-fertilizes with North American transcendentalist philosophers, including Ralph Waldo Emerson and Henry David Thoreau. This tradition continues, albeit in evolved and fragmented ways, through John Muir and current day **deep ecology**.

Thoreau (1817–62) is widely celebrated as an early environmentalist, with a distinctly American reverence for wilderness. When he surveyed a forest, he saw it through very different eyes to Gifford Pinchot, developing a discourse which included a greatly expanded sense of nature's value. In his unfinished manuscript, *Wild Fruits*, he describes remaining forests in New England as

Extractive		Non-extractive				Futurity
Timber products	Non-timber products	Research	Recreational	Spiritual and aesthetic	Ecological functions	Option use and bequest value
lumber wood-pulp	foods resins dyes animal fodder grasses green manure	scientific research education	park visits bird-watching hunting hiking canoeing tourism	aesthetic improvement scenic value place of retreat place of meditation self-realization cultural landscapes of indigenous peoples	soil-conservation carbon-fixing biological diversity water conservation microclimate regulation regional climate regulation	potential use by current generation assumed needs of future generations

Figure 10.1   The instrumental values of forests

'little oases of wildness in the desert of our civilisation' and professes to feel 'something akin to reverence' for them (Thoreau, 2000, p. 168). Thoreau believed that human quality of life is improved through familiarity with wild areas and he lobbied for the preservation of remaining ancient forests. It is a tradition of thought, evident also in the English Romantics such as Wordsworth, that human encounters with 'wild' landscapes bring spiritual self-development. This preservationist argument, based on a reverence for wild nature and a sense of its significance for human well-being, was taken up by John Muir, perhaps the most influential figure in North American environmentalism.

The Romantic and transcendentalist movements provided a radical, but unsuccessful, challenge to modernity, rejecting its reductionist and dualist constructions of nature. The deep ecology movement, sometimes rather confusingly described as 'New Romantics', has built upon this challenge to modernity, becoming an important basis for determining alternative values and objectives for forestry. The deep ecologists' expanded sense of the value of preserved nature, both for human well-being and in its own right, has captured the popular imagination, striking a chord with many individuals' gut-felt or religious sense of reverence for nature.

Despite a broad spectrum of deep ecological thinking, there are some fundamental values that are widely asserted. Most importantly, deep ecologists reject dualism, believing, in the most profound way, that humans are a part of nature. This basic tenet has important implications. Firstly, if we are profoundly part of nature, one can argue from a sense of self-interest, that nature should not be unnecessarily injured: to harm nature is to harm oneself (Pratt, Howarth and Brady, 2000). Secondly, if all things are so profoundly unified, value can only be prescribed to the whole and parts of that whole cannot be more or less valuable than each other. It makes no sense to say that one part of nature is more valuable than another, just as it would make no sense to claim that one part of a circle's perimeter were more important than another. By such reason, a human being cannot be more valuable than a grizzly bear which, in turn, cannot be more valuable than a leech. As a result, deep ecologists believe in biocentric egalitarianism, the view that all parts of nature have an intrinsic and equal value, and an equal right to existence.

As the introduction to this book suggested, the rejection of dualism is not an easy change of discourse for those raised in a Western intellectual tradition. For this reason, the deep ecologist programme is founded on the personal quest for spiritual development or self-realization. Rather than prescribe a detailed set of beliefs to which deep ecologists should subscribe, leading protagonists such as Arne Naess have simply invited their readers to reflect on their own relationship with the rest of nature. Such reflection can take any number of forms although the essential ambition is a progressive recognition of a oneness with nature. The term 'self-realization' is particularly apt in this context because the object of meditation is to recognize one's self in the rest of nature and, perhaps, to recognize the rest of nature in one's self. This entails a 'realization' of an expanded sense of 'self', not only including one's own mind and body, but, ultimately, the entirety of nature.

Spiritual growth, or unfolding, begins when we cease to understand ourselves as isolated and narrow competing egos and begin to identify with other humans from our family and friends to, eventually, our species. But the deep ecology sense of self requires a further maturity and growth, an identification which goes beyond humanity to include the non-human world (Devall and Sessions, 1999, p. 201).

The immediate benefits of such self-realization are clearly personal and spiritual. However, the aggregate of individual change, with the new values this entails, will amount to a social movement with strong convictions. Thinking back to the mode of production model of society (p. 44), this is an idealist approach to social development, based on a view that change is initiated through a change of consciousness (in the realm of cosmology), leading to subsequent changes to the forces and relations of production. To take a simple example, the new deep ecological consciousness will, if it effects sufficient numbers, result in policies that preserve existing forests. At the personal level, deep ecological experiences can clearly motivate environmental preservation. As Gottlieb (1995, p. 9) explains, 'As we sense our continuity with leaf, stream, and butterfly, we manifest a global or ecological consciousness far from the domineering and consumptive obsessions of modernity'. Of recent fame, Julia Butterfly (Julia Hill) underwent a spiritual experience through an encounter with California's west coast redwood forests and subsequently spent two years up a tree, putting her life in grave danger in order to defy loggers. Pepper (1996) states that deep ecologists choose their values before they choose their science (you might consider whether they are unique in this). The values of deep ecology tend to favour adoption of the new sciences of complexity, and particularly the radical holistic credentials of Gaianism.

Despite widespread popular appeal, deep ecology has failed to win over many academics within environmental studies disciplines. At one level there is detailed concern about philosophical robustness and, at another, a general sense that deep ecology amounts to little more than a resurgence of populist folk mythology. Evernden (1992) refers to the 'nature-as-self' project as a form of self-indulgence in which, contrary to stated intent, the individual self is promoted to new heights of importance. Pepper (1995) is among many who find it hard to see how a supposedly non-political movement can realize the protection of both humans and non-human nature, especially where the sources of learning and inspiration are the sites of least human occupation (i.e. wilderness). The social credentials of deep ecology are also marred by perceptions of elitism, especially relating to the idea that those with access to 'wild' environments are advantaged in the quest for wisdom. If there is no division between humanity and nature, why not learn from other examples of nature: the city park, the backyard or even town centres? Mason (1999) is worried about the fundamentalism of the basic tenets of deep ecology, which he sees as an obstacle to progressive communication with other discourses. We ended the previous chapter by ourselves suggesting that communication between discourses is a vital process for environmental management. Many others have simply struggled to see how biological egalitarianism can serve as a practical guide to 'what to do'. From what has preceded in this book, it will be clear that we are not advocating deep ecological positions here. However, we do promote the need to empathize with alternative viewpoints and recognize their valuable contribution to past and future democratic processes.

10.4 Case-study: British Columbia

The Canadian province of British Columbia covers 95 million hectares, 64 per cent of which is forest, contributing around one-third of Canada's forest products by both volume and economic value (CFS, 1999). Some 92 per cent of British Columbia's logging is carried out by clearcutting, practically all of which is in ancient 'old growth' forests. Clearcutting involves the total removal of tree cover from an area, and this has placed the Ministry of Forests at loggerheads with environmentalists and First Nation (aboriginal) groups. Replacement plantations require at least 80 years to mature sufficiently for a second harvest and, since most of these plantations are less than 30 years old, logging companies continue to erode the frontier of remaining ancient forests. Recent conflicts centre on the rare temperate rainforests of the west coast, including Vancouver Island and the 'Great Bear Rainforest' (Figure 10.2). More than one-quarter of the world's remaining temperate rainforest is in British Columbia, much of which is threatened with removal. Of the province's 353 rainforest valleys, 80 per cent have already been developed for forestry, and much of the remainder is subject to development plans (Greenpeace, 1998). Colleen McCrory, founder of the Valhalla Wilderness Society, has labelled British Columbia as the 'Brazil of the North' (Rowell, 1996), but despite such criticism, the federal and provincial governments have continued to support the logging industry because of its contribution to the economy. Canada exports around $40 billion worth of timber products annually, $13.2 billion dollars of which came from British Columbia exports in 1998 (CFS, 1999). The main products are softwood lumber, paper, wood pulp and newsprint, with the major markets in the United States, EU and Japan. In British Columbia one in 11 of the workforce are directly or indirectly employed in forestry (CFS, 1999).

North American forest conflicts are often presented as incidents of the 'wise use' discourse of Pinchot pitched against the 'preservation' discourse of deep ecology. This is a helpful simplification, although the reality is rather more fragmented. The logging dispute is a highly complex environmental conflict, owing to the variety of different **stakeholders**: individuals and groups with an interest in how forests are managed. In postmodern environments, the involvement of all stakeholders is desirable on moral and pragmatic grounds. It is morally desirable for people to be democratically included and for their particular viewpoints to be recognized. It is pragmatic because the exclusion of major interested parties is ultimately dysfunctional and unsustainable, leading to expensive conflicts, the erosion of social capital, failure to incorporate important local knowledge and lost opportunities for innovative negotiation and partnerships. These and other arguments for inclusive forms of governance

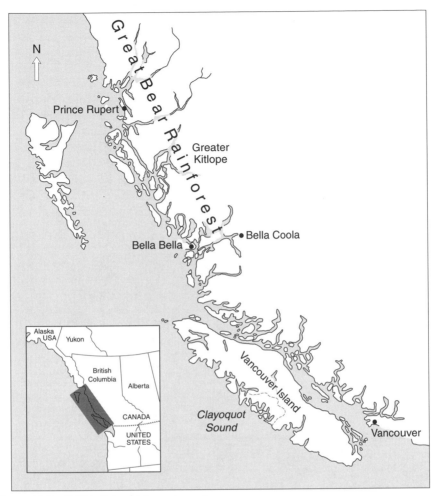

Figure 10.2   British Columbia's west coast

are now so widely accepted that 'participation' has been described as the 'new orthodoxy' (Jeffery and Sundar, 1999). In one sense, the state is making a partial withdrawal, delegating decision-making to both local and supranational institutions. However, such withdrawal in itself creates a new and difficult function. In British Columbia the state must construct institutional mechanisms that will enable the following stakeholders to partake in forms of ecological democracy:

First Nations. The priorities of First Nation groups are aboriginal rights and land titles, pursued through litigation and negotiation. Many territories in British Columbia have never been ceded and in 1997, the Supreme Court of Canada found that British Columbia had never had the legal right to extinguish aboriginal rights and land titles (Mason, 1999), and that lands subject to

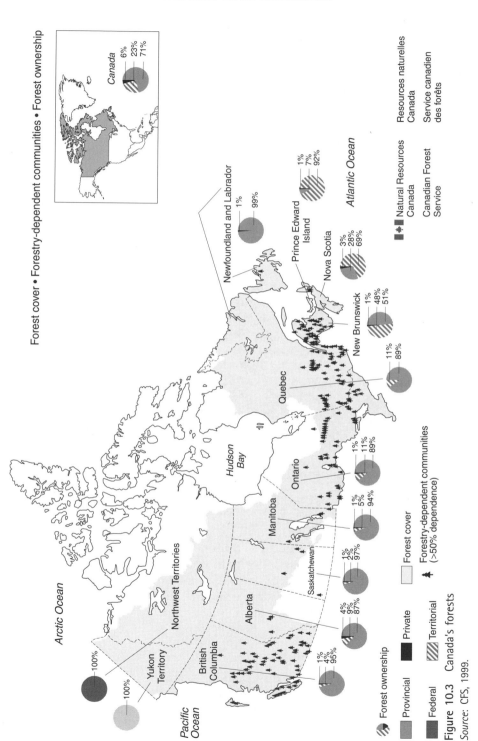

Figure 10.3 Canada's forests

Source: CFS, 1999.

aboriginal claims should not be used in ways that are irreconcilable with their interests (CFS, 1999). Claims for aboriginal rights centre on the reinstatement of forest-based practices that were customary prior to colonization. The granting of aboriginal title provides the right to exclusive use and occupation of forest lands, not restricted to customary practices. At present, 51 First Nations in British Columbia are negotiating treaties, and they are mostly against clear-cutting of forests on disputed lands. Following tenure resolution, clearcutting is not ruled out by all communities. As a generalization, First Nations tend not to be against all logging in old growth forests, providing that they are empowered to control it and benefit from it.

Social ecologists. Drawing on a tradition of North American anarchism, especially the works of Murray Bookchin, social ecologists are materialists, believing (contrary to deep ecologists) that the material (rather than the spiritual) causes of environmental problems are the first priority for reform. These material concerns centre on issues of social justice relating to the control and ownership of resources. There is therefore an allegiance with First Nation groups, to the extent that both prioritize tenure reform leading to local self-determination and local control over resources.

Deep ecologists. Deep ecologists are mostly opposed to all logging of old growth forests. In Canada the deep ecology movement is not divorced from more politically and socially orientated movements and many proponents are also involved in, for example, issues of social justice relating to First Nations (Mason, 1999). However, such alliances may prove to be fragile because, as mentioned, First Nations do not presume against logging. Protests have frequently led to arrests and even violent responses from loggers. Deep ecological sentiments have a popular appeal in Canada. For example, Wagner *et al.*'s (1998) survey of public opinion in Ontario province revealed a mean response of 'agreement' with the statement 'all species, including humans, have an equal right to co-exist on the planet'.

Environmental Non-Governmental Organizations. The Western Canada Wilderness Committee, Valhalla Wilderness Society, Forest Action Network, Raincoast Conservation Society and Greenpeace Canada are among many campaigning groups seeking to protect old growth forests. While member-ship is diverse, it is interesting to note that their campaigning materials have dealt extensively with issues of employment, thus addressing the common charge that environmentalists are 'anti-jobs'. For example, Greenpeace's 1998 publication *British Columbia Communities at the Crossroads* seeks to trans-cend the old 'jobs versus the environment' mindset and presents research demonstrating that social and ecological sustainability are mutually reinfor-cing. The current system of scientific forestry (clearcutting followed by plantation) is generally seen as a giant experiment that amounts to organized irresponsibility.

Local communities. The difficulty in generalizing about local community opinions is exacerbated by the fact that every other grouping presumes to speak on their behalf. In reality, whether we are talking about a village, city or otherwise, it is important to avoid mistaking a place as holding a single community, a term that is laden with assumptions of shared values and priorities. Within British Columbia, settlements may combine people employed in the logging industry alongside those employed in high-tech industries that have migrated to the area because of the quality of the environment. According to a 1997 survey carried out by the Ministry of Forests, 77 per cent of British Columbia's residents desire an end to clearcutting (Greenpeace, 1998).

Logging companies. The major logging corporations in British Columbia are MacMillan Bloedel, Western Forest Products (WFP), International Forest Products (Interfor) and West Fraser Timber. Loggers operate, in principle, according to scientific calculations of sustainable harvests. They see no practical alternative to logging old growth forests, but have been more willing to consider alternatives to clearcutting, based on systems of selective extraction. Logging companies stress that the region's high standards of living were built on this industry and must be maintained by it.

Wise-use movements. These are 'Pinchotist' groups who rally around calls for sustained employment and economic development based on increased access to and 'wise use' of resources. Such groups present themselves as community-based although Rowell (1996) details how they are often funded and orchestrated by logging companies. Wise-use groups are often hostile to environmentalists, seeking to undermine them by branding them as 'hippies', 'dope smokers', 'leeches' and 'welfare bums' (Rowell, 1996, p. 185). The umbrella group, the 'Forest Alliance' was formed in 1991 and continues to lobby widely for logging interests, employing a rather more conciliatory tone.

Consumers. In addition to targeting the logging industry itself, environmental NGOs are seeking to eliminate their customers. This has met with considerable success on both sides of the Atlantic and the market place, driven by fear of consumer boycotts, could yet be the force that demands the end to logging in old growth forests. Consumer led environmentalism has been aided by recent drives to certify products that are sustainably produced. The most sophisticated of certification services is offered by the Forest Stewardship Council (FSC) whose principles embrace a broad set of criteria for sustainability that include requirements for inclusive stakeholder dialogue and the recognition of aboriginal rights to 'own, use and manage their lands, territories and resources' (FSC, 2000, Principle 3). Only small areas of British Columbia's forestry have been certified although WFP's operations are currently being audited (Forest Alliance, 2000). If WFP gains certification of its old growth logging, environmentalists and First Nations will be outraged. FSC principle 9 states that 'Management activities in high conservation value forests shall maintain or enhance the attributes which define such forests' (FSC, 2000). It is hard to envisage how any logging could enhance or maintain the attributes of ancient rainforest.

International communities. In addition to local and regional stakeholders, there are many supranational institutions that represent global interests in forest protection.

10.4.1 Institutional basis of forestry

The existing system of tenure in British Columbia is dominated by the provincial government's legal ownership of 95 per cent of forest resources (some of which is disputed by First Nations). Forest management is under the jurisdiction of the provincial government who enter into forest-management agreements with industrial loggers. Logging companies hold agreements over large territories: for example, WFP controls 680,000 hectares in the Great Bear rainforest. Companies produce forest development plans and are issued with licences to extract prescribed volumes of timber according to practices dictated by the British Columbian Forest Practices Code. Tree Farm Licences are subject to long term harvest level (LTHL) calculations set by the government's chief forester. The LTHL is supposed to be a harvest level that can be 'maintained indefinitely' and is based on the assumed growth rates in replacement plantations. This is used to determine the 'annual allowable cut' (AAC) which is 'the basis for regulating harvest levels to ensure a sustainable supply of timber' (CFS, 1996).

The all-important process of setting values and objectives is concentrated in the corporate sector and tends to be at odds with the values and objectives of other genuine stakeholders. In the early 1990s the prioritization of the extractive values of forest resources led to a high profile protest over Clayoquot Sound, a 350,000 hectare reserve of ancient forest on the west coast of Vancouver Island. Licences for clearcutting were issued to MacMillan Bloedel and Interfor (Rowell, 1996). In the summer of 1993, 800 people were arrested during direct actions to prevent logging and, partly as a compromise, 35 per cent of the area was provided with protected designation. First Nation and environmental groups were far from satisfied, including the Nuv Chah-Nulth people who had never even ceded this territory to the state (Rowell, 1996). The protest took on an international dimension when environmental NGOs persuaded major European pulp consumers (including Kleenex and Scott) to withdraw their custom. In the United States, the *New York Times* did likewise. In 1994, Premier Harcourt felt it necessary to visit Europe to defend the government's forest management policies; in the same year the government began funding the Forest Alliance. In May 1995 the independent Clayoquot Sound Scientific Panel recommended an end to clearcutting and proposed 'variable retention silvicultural systems', a form of selective tree extraction that is intended to leave key ecosystem functions intact.

Following a long and bitter campaign on all sides, Clayoquot Sound was designated as a UNESCO Biosphere Reserve on 5 May 2000. Prime Minister Chrétien enthused about 'the spirit of partnership and cooperation that brought together First Nations, local communities and governments to achieve this designation' (Government of British Columbia, 2000). The reality was more of a spirit of conflict and the years of ill-will between opponents will not be easy to overcome. The biosphere reserve designation carries no legal authority and is effectively a voluntary commitment to sustainable management: 'The Clayoquot Land Use Decision which sparked the mass protests of 1993 and successive years still holds' (Friends of Clayoquot Sound, 2000a). Despite reservations about the legal status, including the absence of legal tenure reforms, the Canadian government has demonstrated a commitment to a new institutional management arrangement that could ensure inclusive dialogue and decision making, what we consider to be a step towards sustainable forestry. The new institutional basis centres on the Clayoquot Biosphere Trust, a not-for-profit organization made up of First Nations and local communities. The Trust works under the mandate of protecting environmental values, promoting sustainable economic development and supporting research, education and training (Government of British Columbia, 2000). Government commitment to the Trust has been underlined by a $12 million endowment which will generate an annual income of around $700,000. Logging will be allowed outside of a core protected area of the reserve.

Testing times lie ahead for the Biosphere Trust, which needs to develop the 'social capital' to manage diverse views. Social capital includes trust between individuals and groups, the evolution of norms, rules and sanctions and the development of processes of reciprocity, exchange and connectedness (Pretty, 1998). Just weeks after Biosphere designation, it became clear that there is a long way to go before Clayoquot stakeholders can work together in a spirit of trust and reciprocity. In May 2000, Interfor resumed logging just outside the core preservation area, leading to renewed direct actions to stop them (Friends of Clayoquot Sound, 2000b). In reality, it would seem that deep ecological arguments will not prevail and compromise is needed. Those seeking preservation outside of the biosphere's core may have to succumb to economic based discourses and enter into partnerships that demonstrate that preservation is an economically attractive alternative. This might involve developing local economies around non-extractive industries such as ecotourism and possibly even initiatives such as carbon trading.

New institutions such as the Biosphere Trust play a role in transforming and developing social capital, thereby transforming the relationships within society and between society and non-human nature. However, new institutions are always in danger of being dominated by old institutions, a lesson learned from Canada's attempt at participatory management introduced through the Commission on Resources and Environment (CORE). Against the backdrop of early 1990s protests, CORE was established to introduce new institutional structures to deal with land use conflicts (Mason, 1999; Rowell, 1996). Participation of various stakeholders was to be institutionalized through 'regional tables' at which land use planning decisions were democratically debated. However, CORE foundered because old sectoral interests prevailed while other voices were marginalized. For one thing, the Ministry of Forests retained the power to decide which areas of land were not subject to regional table debate and they used this power to exclude Clayoquot Sound. Eventually, the Ministry of Forests reasserted its dominance and CORE was scrapped (Mason, 1999). New forms of governance do not easily replace old forms of government. Indeed, the establishment of a new institutional arrangement should be viewed as the beginning of a process, rather than as a finished solution. Even if this process is successful, the shift to sustainability in Clayoquot Sound will not be easily replicable elsewhere: the 'scaling-up' of local governance is notoriously difficult, as the case study of Indian forestry will demonstrate. However, the call for such 'postmodern' solutions is spreading. New battle lines have been drawn up by environmentalists, First Nations and local communities, with recent protests over logging in Saltspring Island, the Upper Elaho Valley and the Ingram-Mooto watershed. The Canadian Forest Service has also been experimenting with joint forest management, community forests and model forests, all of which involve the institutionalization of community involvement.

10.5 Forestry in India

In India colonial rule witnessed the erosion of community tenure and the development of forestry as a means of supplying the needs of empire. Following independence in 1947, the Indian government did not institute forest tenure reforms, believing that forest resources should predominantly serve national needs for agricultural and industrial modernization. This thinking was enshrined within the 1952 forest policy:

> Use by the village community should in no event be permitted at the cost of national interests. The accident of a village being situated close to a forest does not prejudice the right of the country as a whole to receive the benefits of a national asset . . .
>
> . . . rights and interests of future generations [should not] be subordinated to the improvidence of the present generation (Government of India,1952, para. 7)

In the 1960s and 1970s 'modernization' formed the cornerstone of international development discourse. The call for a modern, scientific and, above all, a productive forestry sector came from domestic bodies as well as international aid organizations. In 1970 the US Agency for International Development urged India 'to replace a significant percentage of the mixed tropical hardwood species with man-made forests of desirable species' (Gadgil and Guha, 1992, p.188). Domestic advice concurred, with the National Commission of Agriculture (NCA) calling for a 'dynamic programme of production forestry':

> Considering the advantages of an aggressive man-made forestry programme, we
> feel that the future production programme should concentrate on clear felling of
> valuable mixed forests, mixed quality forests and inaccessible hardwood forests and
> planting with suitable fast-growing species yielding higher returns per unit area
> (NCA, 1972, cited in Kumar, 1992, p. 21).

As we noted at the beginning of this chapter, such 'scientific forestry' has considerable implications for forest-dependent communities, because homogenization of forests requires the homogenization of systems of control and management. Furthermore, the scales and rhythms of mass plantation forestry were often at odds with local needs for forest access; for example, a large plantation can cut villagers off from the old forests that they are dependent upon for basic needs. Thus, while local communities were further alienated from forest management, they were also forced into continued conflict with the forest department, often depending on activities that are technically illegal in order to provide household requirements for fuel wood, construction timbers, fencing materials, animal fodder, green manure and other non-timber forest products. Partly as a result of this continued breakdown of relations between state and villagers, timber smugglers and poachers have prospered, forging profitable alliances with the elite on both sides.

Between 1947 and 1977, India's forest cover decreased from 40 to 20 per cent, leaving only 11 per cent of the land with adequate tree cover (Fernandes, 1987). Despite a longstanding policy to bring forest levels up to 33 per cent (60 per cent in mountainous areas), degradation and deforestation have continued. Even in some of the more forested areas of India, the loss of forest is causing increasing hardship. In most parts of rural India, more than 90 per cent of households still rely on biofuels (trees, dung, crop residues) as their primary energy source (Agarwal, 1998). At the sharp end of resource depletion are the women responsible for collecting fuelwood, who are having to travel increasing distances and having to switch to lower order fuels such as twigs, leaves or cow-dung. In many villages, farmers also report changes in microclimate, the drying of wells and streams during the summer months and increased incidence of flooding during the monsoon. Compared with British Columbia, deforestation has more immediate, concrete and hazardous implications for local livelihoods.

Conventional wisdom, such as that advocated in Garett Hardin's (1968) influential work on 'the tragedy of the commons' (cf. p. 133) suggests that there are two main solutions to managing environmental resources for which access is difficult to limit: they can be privatized or they can be nationalized. These tenurial arrangements were seen as the only viable ways of ensuring that such resources were not treated as 'open access', leading to a disastrous free-for-all. However, observation and experience have found Hardin's thesis wanting. The nationalization of forest resources has led to the alienation of local people and the demise of traditional systems of management (Guha, 1983); privatization has often impoverished the majority while enabling short-term gain for the few whose decision making is distanced from the consideration of environmental impacts (Jodha, 1986).

Furthermore, as Elinor Ostrom (1990) concluded from a widespread study of **common property resources**, some of the most successful systems of management rely on neither of these dominant solutions. Rather, they operate through forms of local citizenship. In India, NGOs had similarly been observing the success of alternative systems of resource management that involved community self-organization. This led to experimentation with community forestry, most famously in Arabari, in the Midnapore district of west Bengal. This experiment was started in the early 1970s and centred on the formation of Forest Protection Committees (FPCs) which helped to protect forests in return for a 25 per cent stake in marketed products as well as other access agreements (Sarin, 1995). By 1992 there were over 2,000 FPCs in West Bengal (IBRAD, 1992; Deb, 1993) with management responsibility for over 250,000 hectares of forest land (SPWD, 1993). Research revealed that 74 per cent of the FPCs were functioning effectively (SPWD, 1992) – a level of success which had to be partly understood within the context of political support from the Marxist government as well as the emergence of a strong *panchayat* system of local government (Saxena, 1993).

From the mid 1970s to late 1980s such empirical experience, linking with the postmodern turn in social science, led to a shift to participatory forest management, initially developed through the social forestry programme. This was largely an effort to encourage villagers to grow trees on farmland, but also included some community woodlots on degraded lands and along communication corridors. Success was geographically limited to areas with readily accessible markets for construction poles and fuelwood. Further, social forestry did not involve any real change in the system of forest tenure, maintaining state ownership and control over all but the most degraded of resources and maintaining the principle that state forests should serve the industrial sector. The 1988 Indian Forest Policy was the first major sign that negotiations between villagers, the state and the international community, were yielding a new direction for Indian forestry. This policy (replacing the 1952 policy) had three important and connected commitments. Firstly, the main priority of state forests was to meet 'the requirements of fuelwood, fodder, minor forest produce and small timber of the rural and tribal populations'. Secondly, while state forests were to meet local needs, industrial needs would have to be met through private and joint sector forestry. Thirdly, in order to change the basis of Indian forestry, there was a call for 'a massive people's movement with the involvement of women' (Government of India, 1988, section 2.1).

The pressure for such a dramatic departure from the 1952 policy had come from a variety of stakeholders, all with their own agendas: international aid agencies, Indian NGOs, village community groups and from the Forest Department itself, who were faced with an increasingly hopeless task of improving India's forest resource. Most significantly perhaps, Indian environmentalism has strong roots in Marxist and Gandhian discourses of social justice. As Guha (1995, p. 48) concludes, 'it is probably fair to say that the life and practice of Gandhi are the single most important influence on the Indian environmental movement'. As such, environmental issues are rarely divorced from efforts to redefine who owns, controls and benefits from environmental resources, and there is a strong movement for self-reliant, self-managing village communities. Perhaps less well known, Gandhi had little time for preservationist discourses such as those espoused by deep ecologists: 'there was nothing of the romantic in Gandhi' (Guha, 1995, p. 60). It is partly due to this cultural heritage that preservationist actions, such as the creation of National Parks, have frequently led to dispute. This has not been due to a lack of concern for the loss of habitat, but due to the treatment of tribal people who traditionally lived in the park areas.

10.5.1 Joint Forest Management

On 1 June 1990 a government circular was issued to all of India's states entitled *Involvement of Village Communities and Voluntary Agencies in the Regeneration of Degraded Forests* (Government of India, 1990). This circular called on state governments to implement a policy of Joint Forest Management (JFM) and, by 1996, an estimated 1.5 million hectares of forest were being jointly managed by 10,000–15,000 community institutions (Sundar, 2000). JFM is a partnership arrangement that provides an alternative to the dominant models of resource management (state control, individual and corporate ownership). The partnership is between the state forest departments and villagers, with an intermediary role for NGOs. Importantly, participation is not left to *ad hoc* arrangements. It is guaranteed through legislation and organized through formally defined institutions, generally known as village forest committees (VFCs). While ownership of forest lands remains with the state, the VFC is granted certain legal entitlements. Firstly, it is entitled to a role in managing dedicated local forests: deciding what trees to plant, where to plant them, how to go about protecting them and so on. Secondly, the products harvested from these lands, mainly poles, timbers and fruits, are typically divided 50 : 50 between the VFC and the Forest Department. Ultimately, the VFC should become a robust and sustainable institution, generating sufficient economic, social and ecological capital to continue with forest development and protection. VFC activities are not restricted to forestry and a proportion of income can generally be spent on village development activities, from subsidizing fuel-efficient cooking stoves and biogas plants to addressing common priorities such as irrigation and drinking water, education, health-care and employment training and generation. The principles at work are reasonably straightforward: by involving villagers in the management of local forests they will benefit from forest products; such long-term flows of benefits will provide the motivation to protect and improve these forests, and therefore the state will also benefit.

JFM has been an ambitious policy, not least because of the sheer scale of implementation: by 2000, it had been taken up by 22 of India's states. We have previously mentioned that replicating successful examples of local governance is difficult and it is not surprising to hear that the process has encountered problems and criticism. For some commentators, the initial strategy of a

Figure 10.4 Bamboo Crafts Workshop, Western Ghats, Karnataka. In some villages in this area, the majority of households are dependent on income obtained from the sale of fuelwood in nearby towns. One of the challenges for Joint Forest Management is to provide an alternative source of income for the women involved in this trade
Source: Adrian Martin.

'partnership' did not go far enough and they would have preferred genuine community forestry in which the state retreated completely. Most, however, see JFM as a step in the right direction. Observations of villages implementing the policy reveal two main categories of problem. The first relates to the legal frameworks for implementation which can be quite restrictive. In Karnataka state, for example, the legal framework prescribed that only forests with less than 25 per cent canopy cover could be included in the scheme, thus restricting community involvement to the most degraded forest lands. This reflected the fact that the Forest Department does not trust villagers to manage valuable stands of forest and, in the words of one senior official, does not see why it should provide villagers with 'ready-made chocolate'. But lack of ready-made chocolate is a problem for institutional sustainability due to the limited opportunity for short- and medium-term returns on their efforts. Exotic tree species, such as acacia, do grow remarkably quickly, potentially achieving a first harvest after eight years – but even this constitutes a long wait for people whose labour is often spent working for their next meal. It is asking a great deal to expect villagers to provide freely of their time and labour over such a long period before they are able to reap the harvest. Perceptions of this income gap are exacerbated by previous experience of plantation forestry, especially where they have been illicitly felled prior to maturity. In February 2000 a new Indian government order extended JFM from 'degraded' forests to include 'better' forest areas – a step towards a realization that JFM will succeed if villagers genuinely benefit from it.

The second category of problem relates to social relations within new institutions. As was identified with the Clayoquot Biosphere Trust, new institutions do not start with a blank slate. They are imposed onto old places, complete with old elites, old discourses and old antagonisms. New institutions reproduce, transform and sometimes even destroy old alliances, hierarchies and agendas. We will cite two examples here, the first relating to the destruction of existing (possibly beneficial) arrangements, the second relating to the reproduction of existing (probably undesirable) arrangements. Firstly, the unworkable system of state control has led to the evolution of coping strategies over the decades. These are informal arrangements that have led to local people having rather more control over local resources than is generally recognized. They take the form of **customary arrangements**, often technically illegal, such as systems of bribery – the forest guard turns a blind eye if you pay him a fine. A rather more favourable, but equally hidden form of custom, is the management role performed by women. In many Indian villages, women are the main users of forest resources, being responsible for fuelwood collection, gathering of non-timber forest products and, less generally, the grazing of cattle. As such, they have been *de facto* forest managers, with the most intimate understanding of the resource. VFCs, which are often dominated by the village's elite males, can threaten such important, informal systems of management. Indeed, when a predominantly male management committee takes on the role of local forest management, deciding which areas to plant, which areas to close to grazing and so on, this can actually lead to the unintended consequence of further marginalization of women. Project planners are realizing that any change to existing systems of resource management and tenure, however benevolent they may appear on paper, can create risks for the poor, the marginalized and women. Furthermore, a **gendered analysis** is always necessary.

The second example, of the perpetuation of old relationships, relates to the difficulty of the forest department adjusting to its new partnership role. As was mentioned in relation to Canada's CORE programme, old power hierarchies can be asserted within new institutions, diminishing the prospects for successful local capacity building. To draw again on examples observed in Karnataka state, the initial planning process is supposed to begin with a two-day **participatory rural appraisal** exercise, in which villagers take part in workshops to map out local resources, define their needs and work towards the production of a management plan for dedicated forest areas. However, this has often been a fairly hollow exercise because the forest department pre-empts the process with important decisions. Frequently, a practice of 'forward planting' is engaged in, which essentially means that VFC plantations are initiated before the VFC has even become involved. This involves choosing which areas of forest are to be included under JFM, which areas are to be planted and what species are to be planted there. The forest department is not entirely to blame for this: they are used to employing their expertise for such matters; they are aware that the sooner trees are planted, the sooner the

Figure 10.5 Participatory rural appraisal as part of the Joint Forest Management process, Kabbe village temple, Karnataka. This photo shows women actively involved in the process. However, cultural constraints, together with busy schedules, make it difficult for women to attend this kind of meeting
Source: Adrian Martin.

Figure 10.6 Woman cutting wood, Western Ghats, Karmataka
Source: Adrian Martin.

benefits can be reaped; and they are under pressure to meet plantation targets. The downside is first and foremost a lost opportunity for genuine involvement and capacity building, but also discontent with the type of trees planted and, in some cases, the choice of locations that interfere with access to other forests.

While problems persist, joint management is a process rather than a product and villagers and foresters are involved in a learning experience that will hopefully, through both conflict and cooperation, develop into mutually desirable working relationships. New forms of governance based on community involvement are now commonplace in all parts of the world, dealing with the planning and management of forests, water, fisheries, new settlements and a host of other local issues. Local Agenda 21 is an important example that will be examined in Chapter 13. With such a groundswell of new institutional arrangements has come a massive research effort, reflected in books and journals, and we are beginning to understand much more about factors that contribute to success and failure. One of the key issues for Indian forestry is how to implement bottom–up local forms of governance, through top–down policies developed by the state. This could be the development paradox of the century and geographers, with their sensitivity to the complex relationships of place, will have an important role to play.

10.6 Further reading

A special edition of the journal *Development and Change* (31, 1) provides a series of studies relating to 'Forest Lives and Struggles', including a study of J. F. M. Mason's (1999) *Environmental Democracy* includes two chapters on forestry in British Columbia.

Ostrom's (1990) *Governing the Commons* remains an important work on tenure and new institutions and Tendler's (1997), *Good Government in the Tropics* provides pertinent success stories relating to new partnerships between the state and civil society.

Extending beyond the bounds of the chapter, you should research the discourses of ecofeminism. Try Plumwood (1993), *Feminism and the Mastery of Nature*, or a number of readings in Smith (1999), *Thinking Through the Environment*.

10.6.1 Websites

You can keep up-to-date on Indian forestry through the journal *Down to Earth*, available online: http://www.cseindia.org/html/dte/dte.htm
and the journal *Economic and Political Weekly*: http://www.epw.org.in/
For British Columbia, Friends of Clayoquot Sound are on: http://www.ancientrainforest.org/ and the Canadian Forest Service on http://www.nrcan.gc.ca

Chapter 11

The postmodern urban environment

11.1 Introduction

Cities are fast becoming the principal human habitat. Eighty per cent of the population of high-income countries live in cities, and at the turn of the millenium around half the world's people were urbanized. Two-thirds of the world's economic production and consumption takes place in the cities of the rich world, and 12 per cent in those of the developing world. If societies around the world are to make the transition to more sustainable forms of development, then sustainable cities are a key priority.

This chapter uses the approach of urban ecology to examine the breakdown of sustainable relations between people and nature in the modern British city and the potential for their reconstruction within emerging forms of postmodern urbanization. Changed forms of social regulation within and beyond the postmodern city are encouraging new forms of urban living with associated changes in urban design, planning and politics. Nature and ecology are becoming more significant in urban development and redevelopment as these processes are increasingly defined, planned, negotiated and legitimated by and with the discourses of environmentalism. The sustainable or 'green' city has become the stated objective of more and more city governments, with local environmental policy regimes shaping how the urban environment is constructed both materially and discursively. Postmodern urban politics requires such regimes to work more closely with civil society, offering new forms of participation and accommodating the demands of new urban social movements.

11.2 Urban ecology

There is no absolute definition of a city. We will follow Haugton and Hunter (1996) by regarding as cities all urban areas that have 'a more or less regular and recognisable agglomeration of buildings and thoroughfares, where people live and work and also engage in many of their social and cultural activities, and a population of at least 10,000 residents' (p. 14). Cities show a great variety of internal form, functions, dynamics, and regional and international contexts. Their environments have natural, built and social components and a provisional definition suggests that 'a sustainable city is one in which its people and businesses continuously endeavour to improve their natural, built and cultural environments at the neighbourhood and regional levels, whilst working in ways which always support the goal of global sustainable development' (p. 27).

Historically, urban development has been fuelled by the exploitation of human and non-human nature and by the efficiency of the urban environment as an arena for capital accumulation and centralization. Within cities enterprises enjoy agglomeration economies of proximity, scale, density and scope. Transport costs are reduced; markets are larger; there is a greater range of services; and shared access to a pool of labour. Urban agglomeration can reduce such environmental costs as those associated with transport and energy, but cities also produce such agglomeration diseconomies as congestion and higher land and labour costs. It is in cities that the second contradiction of capitalist development is most severely experienced, especially by the poor. Failure to reproduce the conditions of production leads to congestion, pollution, poor housing, ill-health, inadequate welfare services and alienation from nature.

Urban ecology approaches the city in dialectical and relational terms, regarding it as a spatial container of nature–society relations (Kipfer, Hartmann and Marino, 1996; Keil and Graham, 1998). Biophysical structures and processes, such as those associated with landforms, climate and vegetation, shape urban form and spatial relations in ways determined by changing social structures and processes, and are in turn shaped by the built environment (Harvey, 1996; Hinchcliffe, 1999). Ecological conditions make urbanization possible and the resulting social, spatial and environmental relations are regulated by institutions in ways that shape continuing urban development. The structuring of urban space or environments is an important material dimension of this regulation, serving to control social practices and people's thought in ways that help temporarily to resolve the contradictions of capitalist urbanization. Social structures, urban space and modes of regulation are, however, products of human agency and subject to contestation and change. Changed modes of regulation may allow the city to develop in more sustainable ways.

11.3 The Fordist city

Chapter 7 outlined the characteristics of the third Kondratieff wave of development in the world economy that lasted from 1945 to around 1975. In the Western liberal democracies, a Fordist model of development involved the state managing capitalism so that an accelerating treadmill of production and consumption (Snaiberg, 1980) met sufficient needs of capital (increased production, sales and profit), labour (rising wages, living standards and welfare), and the state (increasing tax revenues) to contain social conflict. Fordism eventually ran up against economic, political and ecological limits to growth (Chapter 9) and its crisis was by then reflected in the urban crisis found in many of the world's cities.

While Fordist development and Fordist cities show wide variations around the world, such cities generally consist of a corporate downtown, or central business district, surrounded by zones of industry and housing, all linked together with roads and other infrastructure. There is a functional hierarchy between and within these cities that reflects the power and influence of private and public corporations. Roads, electricity and telecommunications concentrate flows of energy, materials and commuters through large factories and offices, but fast and efficient transport also allows the decentralization of industry and services, suburban living and the reduction of urban densities. The motor car, more than any other technology, has reshaped the twentieth-century city, allowing more and more families to live in the suburbs where they can domesticate nature and enjoy the space that increased consumerism requires. As factories, services and the middle and upper classes move from the inner city it faces a spiral of decline with increasing problems of environmental degradation and social exclusion.

Urban and regional planning and management were key aspects of the Fordist model. They reflected mechanical materialism and instrumental reasoning (Chapter 2), with technocrats regarding the city as a machine and people as 'cogs in the machine' (Mumford, 1961). Planners, architects and traffic engineers, supported by local politicians, sought to deliver new development, buildings and roads with little attention to people's real needs as revealed through genuine public participation. Redevelopment often destroyed communities displaying high levels of mutual aid. The zoning of land use, while separating people from pollution, was generally manipulated by powerful land-developers, residents and politicians in ways that destroyed the ordered chaos of mixed neighbourhoods and created unsustainable social 'monocultures' (Jacobs, 1961). The poor design of buildings and spaces, evident in many tower blocks, housing estates and shopping centres, created such environmental problems as those associated with damp cold housing, and isolated people from one another and the rest of nature.

The Fordist city is dysfunctional and disempowering for many of its residents, particularly those forced to live in the most unattractive and unhealthy environments. **Urban disadvantage** is distributed in complex ways across many social categories (e.g. class, race, gender, physical disability, mental illness, homelessness) and may be multilayered (e.g. black unemployed youth). The environmental and social costs of urbanization fall disproportionately on women since urban planning has codified and prioritized land use according to male values and needs. The zoning of workplaces and housing has made it difficult for women to combine paid work with unpaid work in the home, and those women without access to a car are further disadvantaged by the decline of public transport and the consolidation of services into larger out-of-town centres. Poor urban design has contributed to parts of cities becoming unsafe for women and children, and many children now have little opportunity for unsupervised outdoor exploration and play.

In the Fordist city most people are alienated from nature and the natural resources and services that sustain their lives. People are less aware of the sources of the energy, food and other products that they consume and of the destinations of their waste. They have less direct contact with plants, animals and 'green' landscapes, and are more protected from extremes of weather and climate. Consumer capitalism compensates them for their sense of loss by further commodifying nature and reinserting it into the city in material and symbolic forms. Nature is available from the park, garden centre, bookshop, travel agent and television documentary, but the forms in which it is experienced and consumed rarely reveal the processes involved in its construction or explain the reasons why it cannot fully compensate for the alienation induced by everyday urban life.

11.4 Ecological footprints

An ecological perspective on cities suggests that as open systems they can only maintain themselves and grow by importing low-entropy energy material from their host environments and exporting high-entropy (degraded) energy material back into those environments. The **ecological footprint** of a city is the area of land (and water) required for the sustainable production of the biophysical resources and services used by a given city population living with specified forms of technology at a defined material standard (Wackernagel and Rees, 1996). It can be thought of as a solar collector or the photosynthetic surface continuously needed to recharge the city's batteries. It provides the high-grade energy and material that continuously replaces that which is degraded.

Fordist development greatly increased the ecological footprints of cities in high-income countries. While the ecologically productive land area available in the world fell from 5.6 to 1.5 hectares per capita between 1900 and 1995, such cities typically have a footprint of between three and seven hectares per capita. London's ecological footprint just for food, forest products and carbon assimilation is around 120 times the area of Greater London and is greater than the entire productive land area of Great Britain. Software is now available to allow households and institutions to calculate and monitor their footprints, and footprint analysis is proving a useful tool in planning more sustainable cities (Carley and Spapens, 1998; Simmons and Chambers, 1998).

Such analysis reminds us that modern technology and trade result in the ecological locations of city regions no longer coinciding with their geographical locations. Therefore Rees (1997) suggests that the sustainable city is an oxymoron since no city or city region can be sustainable on its own, but only if its global hinterland is also sustainable. Capitalist technology and trade expand the scope and efficiency of the city's exploitation of the biosphere, accelerating natural capital depletion, reducing global carrying capacity and increasing long-term risk. Sustainability requires city governments in the North to rebalance localization and globalization: to seek greater self-reliance on local rehabilitated natural capital stocks while supporting a more just international economic order that allows governments in the South to realize sustainability on their own terms (Hines, 2000).

As centres of population and consumption, with considerable agglomeration economies, cities have considerable leverage to reduce their footprints and move towards sustainability. **Urban planning and design**, together with changes in people's patterns of consumption, can minimize the city's use of energy, materials and land. Green areas within cities can be used to produce food, regulate climate, absorb carbon, recycle waste, and meet people's aesthetic and spiritual needs for contact with nature. Such areas should be integrated with others in more sustainable neighbourhoods to increase the city's self-reliance and reduce the ecological load it imposes on distant places and the global commons (Trainer, 1998). City development should ideally be zero-impact development. Biophysical resources and services that are destroyed should be reconstructed and so compensated elsewhere.

11.5 Fordist local government and urban planning in the United Kingdom

In Britain the **local state** and elected local government were key elements of the Keynesian welfare state, providing such important elements of the social wage as housing, education and land-use planning, that helped to underwrite mass consumption. Such services were generally unprofitable for the private sector and were provided by hierarchical, bureaucratic and corporatist public institutions that reflected Fordism's organizational form. **Local government** shaped many sites of regulation in the manner suggested by Figure 11.1, and

Sites of regulation	British local governance in Fordism	New developments
Financial regime	Keynesian	monetarist
Organizational structure of local governance	centralized service, delivery authorities; pre-eminence of formal, elected local government	wide variety of service-providers; multiplicity of agencies of local governance
Management	hierarchical; centralized; bureaucratic	devolved 'flat' hierarchies; performance-driven
Local labour markets	regulated; segmented by skill	deregulated; dual labour market
Labour process	technologically undeveloped; labour-intensive; productivity increases difficult	technologically dynamic (information based); capital-intensive; productivity increases possible
Forms of consumption	universal collective rights	targeted individualized 'contracts'
Nature of services provided	to meet local needs; expandable	to meet statutory obligations; constrained
Ideology	social democratic	neoliberal
Key discourse	technocratic/managerialist	entrepreneurial/enabling
Political form	corporatist	neocorporatist (labour excluded)
Economic goals	promotion of full employment; economic modernization based on technical advance and public investment	promotion of private profit; economic modernization based on low-wage, low-skill, 'flexible' economy
Social goals	progressive redistribution/ social justice	privatized consumption/active citizenship

Figure 11.1 New developments in British local governance
Source: Goodwin and Painter, 1996.

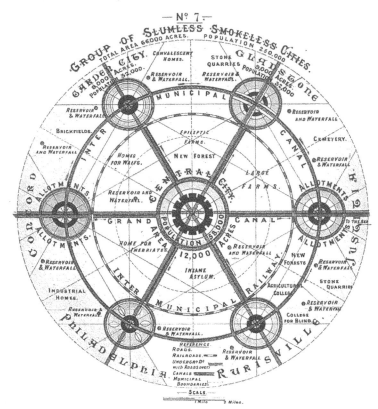

Figure 11.2 The social city
Source: The Town and Country Planning Association, 1993.

together with the policies of national government, its policies on local economic development and land-use planning, sought to prevent uneven development and encourage social and geographical coherence. The planning agenda between 1947 and 1976 was concerned with the redevelopment of congested urban areas and the decentralization and dispersal of industries and population from such areas. The Town and Country Planning Act of 1947 provided a system for guiding and controlling development to conform with approved plans, with county councils as the principal planning authorities (DoE, 1996).

Approaches to urban planning have variously attempted to return the city to nature through deconcentration and dispersal (Kropotkin, Morris, Lloyd Wright); tame nature with technology through dense urban concentration (le Corbusier, Buckminster, Fuller); or strike a balance between the city and nature in city or bioregions (Howard, Geddes, Lynch, McHarg). In his seminal text *Tomorrow: a peaceful path to real reform*, Ebenezer Howard (1889) included a diagram (Figure 11.2) of the social city, a cluster of sustainable garden cities connected by a rapid transport system. The **social city** was the physical realization of Howard's third magnet (Figure 11.3) having the advantages of both the town and the country with none of their disadvantages. Its

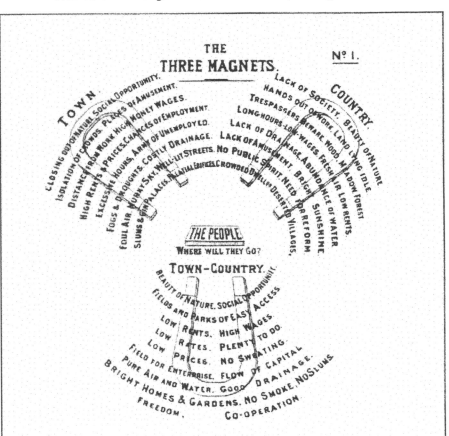

Figure 11.3 Howard's three magnets
Source: The Town and Country Planning Association, 1998.

constituent garden cities were to be totally new towns in the middle of the countryside, surrounded by green belts with farms, and having central public gardens and larger public parks. Howard's social city reflected sustainability goals by greening the city, reducing journey times, recycling organic wastes and providing public facilities, but he was insufficiently aware of the need to conserve biodiversity and other natural resources, and to plan cities in ways that reduced their global footprint (Hebbert, 1998).

The first **garden city** was built at Letchworth in 1903 and others followed in the mid-Hertfordshire corridor. The Town and County Planning Association (TCPA) was formed in the late nineteenth century to challenge the waste and greed of big cities and to campaign for an alternative way of living and working in decentralized small-scale communities in balance with the natural environment. Like the garden city movement, the TCPA advocated an organized exodus of people from overcrowded and unhealthy cities; model towns laid out on scientific lines for health and efficiency; and the ideal of community landowner-ship. These aims were partly realized with the passing of the 1947 Act but the land-use planning system subsequently proved too narrow in scope to promote sustainability (Cullingworth, 1996). It also failed to capture for the commun-ity any significant part of the profits from the increase in land values that arise from development following the granting of planning permission, and so failed to reflect the ideal of community landownership. This encouraged unsustainable development as green-field sites yielded a greater increase in land values than brown-field sites. The state was also denied revenue that could be used to mitigate the environmental impacts of development; for example by planting enough trees to absorb the extra carbon dioxide produced.

Nevertheless, the garden city movement had a profound effect on British land-use in the twentieth century, shaping successive waves of new towns and planned town expansions, and often ameliorating the worst excesses of the Fordist city. At their best new towns provided space, order and openness in good economic and social environments with a full range of services and full employment. They were not, however, integrated with the surrounding coun-tryside, often introduced nature into the city without reference to the needs and participation of residents, and were too ready to accommodate the car. The new city of Milton Keynes illustrates these contradictions (Box 11.1).

Box 11.1   Milton Keynes: city of trees and motor cars

Milton Keynes, the new city in north Buckinghamshire, is a product of the growth pressures that face southeast England, private-sector investment and central government's regional, green-belt, new-town and transport policies. Originating in a plan by Buckinghamshire County Council to avoid overspill in the south of the county, the new city was designated in 1967. It was planned and developed until 1992 by the Milton Keynes Development Corporation which, although formally responsible to central government, proudly operated as an independent development agency dominated by unelected agents.

The planners of Milton Keynes saw it as a futuristic city of the post-industrial (postmodern) era based largely on the new information economy. They maintained some commitment to the ideals of the garden city movement, but thought that the plans for existing new towns had started with too much detail and had proved obsolete before they were realized. Much influenced by European readings of Los Angeles, the grand plan of 1970 envisaged low-rise development of housing, employment and service centres, at low density within a grid system of roads. Milton Keynes would be a flexibile city, keeping its future options open while providing freedom of choice for a highly mobile car-owning society. A grid-based template, surrounded by landscaping, would allow developers to create a postmodern city within broad guidelines.

Crucial to the success of the plan was the material and discursive construc-tion of nature within the new city (Figure 11.4). Milton Keynes is built on

(continued)

(continued)

(continued)

(continued)

(continued)

(continued)

Figure 11.4   Images of nature in Milton Keynes
Source: John Huckle.

low-grade agricultural land with existing settlements at the corners of a designated area that is cut through by the main canal, rail and motorway routes
linking London and the Midlands. Two valleys, separated by a central ridge,
contain artificial lakes to manage surface water, and these also contain linear
parks linked by a further park through the city centre. Paths through the parks
and the rest of the city (green ways and red ways) allow pedestrians and
cyclists to travel through the city in environments dominated by planting and
separated from the main grid roads. Residents living within the grid squares
are screened from these roads by earth-mounding and tree-planting, and
different dominant species are used in different parts of the city. Fourteen
million trees had been planted in Milton Keynes by the early 1990s, and the
city's landscape architects claimed that landscaping provided coherence and
linkage, and was the prime indicator of the 'grand suburban ideal'.

(continued)

(continued)

In television and magazine advertisements, and in other promotional materials, the image-makers of Milton Keynes constructed a city in the countryside, a city close to nature. A television commercial featured an early morning cycle ride through manicured fields to easily accessible fishing, while a brochure described Milton Keynes as 'a quilt of secluded but connected villages'. Comforting images of an imagined past and a stable English suburban society were combined with futuristic images of the city centre and new industry. This masked the reality of poverty and alienation experienced by many of the residents and reinforced a regional ethnicity that remains predominantly white.

Transport in Milton Keynes illustrates its failure to develop as a sustainable postmodern city with a small ecological footprint. An early plan suggested a city based on a monorail loop with high-density development around each station. When this was abandoned planners introduced an innovative 'dial a bus' service but this was soon withdrawn on cost grounds. The deregulation of bus services in the 1980s and the subsequent introduction of street shuttles improved public transport but Milton Keynes remains dominated by the private car. While the grid road system eases traffic congestion and improves fuel efficiency, there is little immediate prospect of the alternative tracked transport system that it was designed to accommodate. Those local centres built in the centre of grid squares do not provide sufficient services or employment to prevent many car journeys, and the secondary grid of walking and cycling routes has not proved as popular as planners hoped. Many residents fail to perceive how these link them to other parts of the city or are deterred by fear of attack on secluded and under-used paths. The shopping centre in central Milton Keynes is far larger than the initial plan proposed and acts as a magnet for the city and the region. The 1991 census revealed that 20 per cent of residents worked outside the area and that there were a similar number of inward commuters.

Sources: Walker, 1982; Bendixson and Platt, 1992; Charlesworth and Cochrane, 1994; Percy, 1996.

11.6 Economic restructuring and competition between cities

Consideration of Milton Keynes reminds us that restructuring of the world economy from the 1970s initiated a new post-Fordist model of capitalist development with attendant possibilities for cities to reinvent themselves in more sustainable forms. The shift to flexible accumulation, using new technologies and labour processes, outlined in Chapter 9, was associated with the intensification of both globalization and localization. Cities attempted to restructure themselves by improving their competitive advantage within the world economy or by developing local economic strategies designed to produce a degree of self-reliance or protection from global economic forces. Harvey (1989b) suggests four forms of competition between cities, and all of these involve the reconstruction of urban nature in various forms.

- Competition within the spatial division of labour. In competing to attract international investment and deter indigenous employers from moving to other locations, cities compete to provide a financial and legislative environment attractive to employers and to provide residential environments attractive for executive living.
- Competition within the spatial division of consumption. Cities compete for a role as consumption centres by initiating and encouraging investment in such facilities as leisure and hotel complexes, marinas, casinos and heritage centres.
- Competition for command functions. Cities compete for world status by seeking to attract the high-order functions of high finance and government.
- Competition for redistribution. Cities compete in the redistribution of financial resources carried out by governments and non-governmental organizations in such forms as regeneration grants and project aid.

While Milton Keynes is one city that has faired well by competing within spatial divisions of labour and consumption and for financial resources from governments and other sources, other cities have faired less well. Restructuring resulted in much urban de-industrialization in Britain, with many factories relocating to green-field sites on the edge of cities or to small rural towns to escape high land values, congestion and pollution. Other industry moved overseas in search of cheap labour, new markets, or less stringent labour and environmental regulation. The local socialism that re-emerged in the 1980s in such areas as Greater London, south Yorkshire and the west Midlands was largely a response to this de-industrialization (Gyford, 1985). Here councils developed their own local economic strategies, supported workers' cooperatives and socially useful production, subsidized public transport, encouraged recycling and energy conservation, and involved tenants and citizens in new forms of democracy and cultural creativity. They attempted more sustainable forms of post-Fordist development, but were abolished by a Conservative government as part of policies designed to restructure local government.

11.7 From local government to local governance

The crisis of Fordism was partly a crisis of the local state and local regulation. The expansion of local government in the 1970s, together with the spread of collective wage-bargaining to the public sector, meant that local government became increasingly costly and bureaucratic. It could not realize productivity gains to offset its rising costs, and citizens were finding its services increasingly remote and unresponsive to their needs. From 1979 successive Conservative governments transformed a uniform system of local government into a more complex one of **local governance**, introducing the new developments shown in Figure 11.1. As Rydin (1999) points out: '[while] government implies hierarchical relationships between tiers of the state, a strong element of top–down control and a firm boundary between the state and outside organisations, governance points to the proliferation of quasi-governmental agencies and a growing formal role for organisations outside the state within the policy process'.

There were four key areas of change (Goodwin and Painter, 1996):

- increasing central control of local finance;
- the privatization and commodification of public services;
- the loss of local state autonomy over the remaining public services; and
- the expansion of non-elected sub-national agencies.

These formed part of a restructuring of the institutions and mechanisms through which local governance operates, and of political structures, projects and alliances. A wider range of actors now has influence over the city economy and environment. It includes the institutions of elected local and national government, but also non-elected organizations of the state as well as institutional and individual actors from outside the formal political arena (e.g. NGOs, private businesses and corporations, supranational institutions such as the EU). Coalitions and partnerships of these actors carry out more flexible forms of policy formulation and implementation, and the resultant **local environmental policy regimes** may be more appropriate for achieving sustainable urban development.

Local governance allows more differentiated spaces of regulation to arise reflecting localized conditions of production and consumption and local constellations of social forces and practices. While Fordism and local government promoted a national welfare strategy that sought to limit social and geographical inequalities, the transition to post-Fordism and local governance under Conservative governments, brought increasing social and spatial inequalities. The market-driven exodus of people from cities to suburbia and the countryside, and from the north to the south, intensified pressure on the green belt and housing market, and contributed to Labour's election victory in 1997. By then the English countryside had become a version of Howard's third magnet on a vast scale, partly through planned decentralization but also through spontaneous, market-driven movement (Hall, 1996; 1998). The challenge now facing politicians and planners was how to reverse this movement by making large cities attractive places to live.

11.8 Post–Fordist local governance and urban planning _____

The shift from Fordism to post-Fordism, or from modernity to postmodernity, offers the prospect of a shift from simple, uniform and unsustainable cities designed and planned against nature to more complex, diverse and sustainable cities designed and planned with nature. In the 1990s international agencies, national and local governments, and NGOs all offered proposals for reorganizing and managing cities to promote sustainability. Some cities became models for the kinds of changes that they envisaged (Giradet, 1996; Satterthwaite, 1999).

In terms of **urban design** there was an increased realization that specifications for houses, workplaces and roads can reduce demand for space heating and transport, and thereby reduce energy demand. Residential areas should be designed to minimize car-dependency, encourage safe walking and cycling, and provide easy access to public transport. A mix of privately owned and rented houses of different sizes should produce socially mixed neighbourhoods that encourage social inclusion, while the addition of local shopping and employment should reduce the need to travel. Buildings should use the most sustainable construction methods, minimizing waste and energy use, recycling land and materials and avoiding harmful materials such as PVC. They should promote water-saving technology and the use of dual water systems, and make optimum use of natural light, ventilation and heating. Plants and animals should be introduced into cities to provide sensory stimulation and connection with nature, and planting should moderate air pollution, noise and extremes of weather and climate. Open spaces should be available for a variety of uses including urban agriculture and community gardening (Hough, 1995; White, 1994).

The shift to local governance encouraged such **'greening' of cities**. At a time when the powers and resources of local governments were being reduced, many used the new agenda to help them redefine their role. From the late 1980s many adopted environmental charters, carried out environmental audits of their activities, produced state of the environment reports and repackaged their services in more 'environmentally friendly' forms. Local governance encouraged such 'greening' with its emphasis on the quality of services and service delivery, and the requirement that local government obtain 'best value' for money spent. It encouraged greater coordination across departments and more cooperation between councils and the private and voluntary sectors. The Local Government Management Board provided councils with guidance on environmental management matters and played a key role in coordinating their response to Agenda 21 (Chapter 13).

After the hostility shown to planning by Conservative governments in the 1980s, it returned to favour in the 1990s. Encouraged by Agenda 21 and the European Commission's Green Paper on the Urban Environment (1991), a report from the TCPA (Blowers, 1993) sought to widen the scope of land-use planning to incorporate sustainability. It suggested that environmental planning should promote the goals of resource conservation, ecological design, environmental quality, social equity and political participation. It should adopt the precautionary principle; reflect the integrated nature of environmental processes and policies (across media, economic sectors and geographical boundaries); and take a strategic approach to decision-making with nested and interdependent environmental plans. These would need to be complemented with reforms of private property rights and liabilities, taxes and charges, subsidies and grants, monitoring and information, and new legislation.

In one of its chapters (Beheny and Rookwood, 1993) the TCPA report updated Ebenezer Howard's ideas by suggesting that the basic unit for the development of appropriate environmental standards should be the social

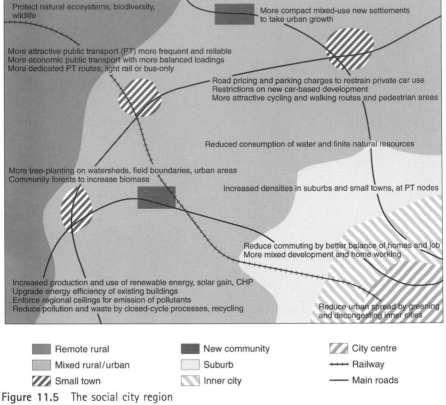

Figure 11.5 The social city region
Source: Beheny and Rookwood (1993).

city region. Within this there should be varying standards (for example for densities, urban form or transport systems) to suit different conditions in, for example, city centres, small towns and new communities, or remote rural areas, but policies should be complementary and contribute to greater sustainability in the region as a whole. This model (Figure 11.5) is more complex and perhaps more realistic than the dominant 'compact city' prescription.

Planning policy did subsequently widen in scope and pay greater attention to sustainability. Governments used planning policy guidance notes on such matters as town centres and retail development to signal shifts in policy and development, plans began to pay greater attention to such matters as the conservation of energy and biodiversity (Selman, 1996). Some structure plans, like those of Hertfordshire County Council, were seen to be at the forefront of environmental planning, testing all the policies they contained against sustainability criteria and indicators (Counsell, 1999a). European legislation required major developments to be subject to environmental assessment, and this process also encouraged the new attention to sustainability. The construction and implementation of local sustainability plans (Chapter 13) further raised the profile of planning and encouraged new ways of involving urban communities and citizens in the planning process (Counsell, 1999b). As with forestry (Chapter 10), postmodern urban planning and design seeks to reinvigorate the multiple voices of local civil society, and gives expression to the discourses of democratic pragmatism, sustainable development and ecological modernization (p. 20).

On coming to power in 1997, the new Labour government unleashed a raft of initiatives to tackle economic and social deprivation and regenerate Britain's cities. These included an Urban Task Force, chaired by the architect Richard Rogers; eight Regional Development Agencies to administer the Single Regeneration Budget and all European regeneration money; the Social Exclusion Unit to ensure that policies on urban renewal were part of an integrated strategy that incorporates policies on health, education and crime;

- One-third of Britain's population lives in its eight largest conurbations. Their population fell by two million between 1961 and 1991 and they continue to lose around 300 people every day to smaller towns and the countryside. This exodus is largely of the well-educated and trained with earned incomes and political power. It has been aided by government policies and the planning system and its main direction is from north to south. Britain's inner cities have high levels of poverty, crime, unemployment and ill health.
- The government predicted the need for 4.4 million new homes between 1991 and 2016. This demand was largely driven by more people choosing to live alone, and presented an opportunity to reinvent urban areas. The government was committed to ensuring that 60 per cent of all new homes in England were built on brownfield sites.
- The government's urban taskforce, chaired by the architect Richard Rogers, called for tough measures to curb the rural drift and help repopulate and revive Britain's cities. These included a national crusade to improve the quality of urban life, create more jobs and recycle an estimated 1.3 million urban buildings lying empty. Ministers were urged to direct more resources to the inner cities and provide a range of fiscal incentives for developers and local councils in special urban priority areas, or low-tax zones. Lord Rogers was disappointed by the government's response, suggesting his recommendations were being sidelined by the Treasury.
- Shelter, the housing charity, claimed that there was a huge unmet demand for affordable housing to rent in southern England. In its view, the new homes being built were too often the wrong homes, in the wrong place, for the wrong people. More 'social' housing, perhaps 40 per cent of planned provision, would have to be built to cope with demand.
- In 1999 government inspectors 'tested' proposals from SERPLAN, a consortium of local councils in the southeast, for a maximum of 914,000 houses in the 12 counties outside London. The inspectors recommended 1.1 million new homes over a shorter period, provoking a storm of protest, but the government adopted a cautious approach with five-yearly reviews and much tougher planning guidelines designed to maximize use of brownfield land and limit urban sprawl. Some 215,000 new houses will be built in the southeast between 2000 and 2005, alongside a further 115,000 in Greater London. Growth will be concentrated along the Thames, at Ashford (in Kent) and Milton Keynes.
- The Town and Country Planning Association accused countryside campaigners of 'deliberate misrepresentation' that might force the government into a policy of 'town cramming' because of impossibly high brownfield targets.

Figure 11.6 Britain's urban crisis: the situation in the late 1990s
Sources: Hetherington, 2000a and 2000b.

and Health, Education and Employment Action Zones in areas of severe deprivation. At the same time Labour continued to reform local governance by establishing a parliament in Scotland and an assembly in Wales, and promoting such innovations as elected executive mayors, cabinet government in town and county halls, referendums on local issues, electronic voting and greater powers for local people over council spending and taxation.

The government hoped that such reforms would counter growing disaffection and apathy with local councils, increase the turnout in local elections, encourage more active forms of citizenship, and so help to contain growing public concern over uneven urban development (Figure 11.6). We will return to an assessment of Labour's policies in Chapter 13. This chapter concludes by considering the increased significance of discursive constructions of nature within the postmodern city.

11.9 Urban citizens and environmental discourse

The greening of local governance and planning, means that urban development and redevelopment are increasingly defined, planned, negotiated and legitimated with the discourses of environmentalism. New houses, business parks and leisure facilities are promoted and sold using images and signs of nature, and urban politics is increasingly shaped by competing environmental discourses. Macnaghten and Urry (1998) explain how public bodies seek to implicate people in sustainable development through a modern discourse that prompts trust in expert systems (the doctrine of naïve environmental realism, Figure 12.2, p. 215). It assumes that citizens are rational agents; that they have faith in science and public policy to identify and manage environmental problems; and that they still believe in the modern project of economic development and social progress. The reality in postmodern times is that these assumptions are no longer valid. There is widespread disaffection or alienation from institutions of science, the media, business and the state, and much pessimism concerning their efficacy or willingness to tackle issues of sustainability. Discussions with focus groups suggest that many people make sense of environmental issues in terms of how they impinge upon their sense of identity. Local, moral and aesthetic meanings of nature and the environment may be more significant than global, instrumental and cognitive meanings, and lack of action may have more to do with a profound mistrust of institutions and politics than a lack of information or awareness.

Urban life has long provided the setting for democratic ideas and experiments with urban politics challenging the agendas of the political class and keeping issues like homelessness, poverty and sustainability in the public eye. Urban citizens run hundreds of thousands of democratic organizations, sustain and build huge amounts of social and organizational capital, and engage in many experiments with local economies or self-managed communities (Box 11.2) that demonstrate that 'better your neighbour localisation' offers a superior quality of life to 'beggar your neighbour globalisation' (Hines, 2000). **Urban social movements** seek to 'democratize democracy' and connect with the wider anti-corporate movement introduced in Box 1.2 (p. 14) and discussed in Chapter 9.

Box 11.2 Exodus

Exodus, is a community living in Luton, 30 miles north of London, that originated as a travelling sound system delivering free music to ravers and advocating the legalization of cannabis. It attracted the attention of the police at a time when a Conservative government was seeking to outlaw outdoor rave parties and crack down on hard and soft drugs. The largely unemployed and homeless members of Exodus squatted in derelict buildings and land in Luton, establishing a housing action zone (HAZ manor) (Figure 11.7) and a

Figure 11.7 Members of Exodus renovate a disused hospital to recreate a housing action zone
Source: Exodus.

(continued)

(continued)

city farm by 'do it ourselves' methods. At HAZ manor they have a communal organic garden, a sustainable water system and are saving for a renewable energy system. They have gradually found an accommodation with the police and Luton Council, and have plans for The Ark, a community centre for others who are socially excluded on Luton's Marsh Farm estate. It will have a non-profit community shop, selling organic fresh vegetables produced by the farm, a wind generator providing energy for the whole estate and cheap entertainment of all sorts for young people. Glen Jenkins of Exodus describes its approach to urban sustainability:

> This is regeneration by the people, for the people. We are taking responsibility for our environment, we want to make it liveable and sustainable. We can't leave it to people who think regeneration is about repainting a few doors and promising computers to schools. What people don't realise is that our philosophy addresses social, environmental and spiritual poverty, as well as problems with money (Saunders, 1999, p. 29).

While many in urban social movements reject the dominant discourse of urban development and seek to establish alternative material and discursive meanings of urban sustainability, geographers such as Harrison and Burgess (1994) have examined the role of discourse in urban environmental politics. Their study of the dispute over a proposed commercial entertainment development on the Rainham Marshes Site of Special Scientific Interest in east London, suggests that distinctive myths of nature are associated with the discourses of developers, conservationists and the media. Developers and conservationists employed different constructions of nature to justify their positions, and local people tried to make sense of their competing claims. They based their judgements on their lived experience of second nature and on utilitarian arguments linked to livelihood rather than abstract arguments based on modern science or the intrinsic value of nature. People's environmental knowledge is the theme of the next chapter.

11.10 Further reading

Each month the *New Internationalist* provides a lively guide to an issue of development and the environment with articles, charts and case-studies. The June 1999 issue looked at green cities. Read the introductory article (Baird, 1999) and others that capture your interest.

In 1995 the architect Richard Rogers gave the BBC radio Reith Lectures. These were subsequently reworked as *Cities for a Small Planet* (Rogers and Gumuchdjian, 1997). Read Chapter 2 on sustainable cities, paying particular attention to the plans for Shanghai and the development of Curitiba in Brazil. Do you find Rogers' arguments about compact sustainable cities, convincing?

Read the chapter on rave culture in McKay (1996) with its references to Exodus. What is the value of such 'cultures of resistance'? Do they prefigure sustainable societies with less emphasis on work and consumerism and greater attention to self-management and community development?

11.10.1 Websites

International Council for Local Environmental Initiatives: http://www.iclei.org
Department of the Environment, Transport and the Regions: http://detr.gov.uk
Environment Resources and Information Centre: http://wmin.ac.uk/eric/
Reclaim the Streets: http://gn.apc.org/rts/
Exodus collective: http://www.exodus.sos.freak.com/index.htm

Chapter 12

Environmental knowledge and the politics of genetically modified food

12.1 Introduction

A glance through the television schedules or popular magazines should be sufficient to remind you that many people in the rich world are preoccupied with food, diet and health. At a time when some would claim that we are healthier than ever before, markets for organic foods, slimming products and therapies of all kinds are growing rapidly. How can we explain these preoccupations and what is the role of environmental knowledge in their production?

One answer is that food has a direct impact on the individual. Whereas most environmental problems are **externalities** with their impacts largely falling on others, unhealthy or polluted food is an **internality** that directly affects the person who eats it. This explains why the market for organic vegetables is greater than that for phosphate-free detergents. In much of the world, food is also a central part of culture, a key element of cultural diversity that people still largely control and are prepared to defend against the kind of attack represented by the spread of fast food.

The significance of food is heightened by the processes of **de-traditionalization** and **individualization** that accelerate with the rise of postmodernity. As traditional authorities, customs and beliefs lose much of their power and influence, people are increasingly required and encouraged to make their own choices and way in the world. Postmodernization leads to the erosion of class, religious and geographical identities, and alongside the old politics of production and wealth introduces the **new politics** of risk and identity. Individuals are less deferential to authority, show less trust in politicians and experts, and increasingly perceive themselves as autonomous in what they can be and do. Many use the expanding range of goods, services and knowledge to develop their aesthetic and cognitive reflexivity, and a further glance at television and magazine advertisements suggests that much of this activity is focused around the reconstruction and representation of human nature or the human body.

While concerns about food, diet and health are clearly shaped by advertising and new forms of production, they are also a product of **risk society**. People are increasingly aware of such risks as those associated with pesticide residues on vegetables, growth hormones in livestock, BSE (bovine spongiform encephalopathy or 'mad cow disease'), hormone-mimicking chemicals (endocrine disrupters), and genetically modified (GM) foods, and the incalculable and irreversible damage these risks may be doing to their health and that of future generations. As we saw in Chapter 9 such risks are associated with the **organized irresponsibility** of governments, corporations and parts of the scientific community, and such irresponsibility contributes to the **fragmentation of science** and growing lack of trust in scientific 'experts'. Risks are generated by the application of scientific knowledge, but science is only beginning to understand some of them and can offer little reassurance. While the new politics forces risk onto business and political agendas, making **risk-management** a key policy field that no company or government can afford to ignore, environmental groups employ new strategies and communication technologies to campaign on risk issues. By challenging governments not to appear partial or biased, to establish independent mechanisms for providing advice, and to adopt more democratic forms of public communication and consultation, they prompt a shift from to more communicative forms of rationality, and ecological democracy and citizenship.

This chapter considers the role of environmental knowledge in the construction, perception and management of risk (Eden, 1998; Lash, Szerszynski and Wynne, 1996). After a theoretical introduction to the role of environmental knowledge in environmental politics, it focuses on the introduction of GM crops in both the UK and India. While resistance to such crops in the UK and Europe focused mainly on possible risks to food safety and the environment, that in India was more concerned with the control of seeds, the pirating of genetic resources, and related issues of dependency, debt and food security. In both countries campaigners critiqued applications of biotechnology based on normal rather than post-normal science and used new information and communication technologies to create networks of support across the world.

12.2 Normal and post-normal science

You will remember that Chapters 1 and 2 introduced the social construction of nature through both material and representational practices and suggested that this is best understood within the philosophical framework provided by critical realism. A key advantage of this framework is that it can accommodate postmodernism's attention to the social construction of knowledge without losing sight of the connections between language and real structures and processes in the biophysical and social worlds, or suggesting that all knowledge is of equal worth (also see Chapter 9). Critical realism underpins **post-normal or postmodern science** and Figure 12.1 explains how this questions the assumptions and applications of **normal or modern science**. It suggests that new technologies and environmental risks challenge dominant technocratic forms of environmental policy-making and so intensify the legitimation crisis of the state. Governments respond by offering new forms of public consultation and participation that can better accommodate post-normal science and prompt new forms of governance and citizenship.

Normal (modern) science	Post-normal (postmodern) science
Empirical data lead to indisputable facts or true conclusions. Uncertainties are tamed by reducing complex systems to their simple elements.	Recognizes uncertainty and a plurality of competing but legitimate perspectives. Quality replaces truth as the organizing principle.
Scientific knowledge is assumed to be qualitatively different from the lay and tacit knowledge of the public. It seeks orthodoxy, replicability and universality.	Scientific knowledge is complemented by non-scientific expertise or elements of the public's lay and tacit knowledge (local knowledge, contextual knowledge and active knowledge). It seeks plurality and heterogeneity but avoids relativism.
Accredited scientific experts discover 'true facts' for the determination of 'good policies'. The public are seen to lack expertise and knowledge of science and are effectively disqualified from participation in scientific debates. Expert scientists speak for the environment in policy debates.	Scientists participate in dialogue with the private sector, government and civil society to assess the quality of scientific knowledge in the context of real-life situations. Scientists help citizens to produce citizens' science and speak for the environment in policy debates.
Abstract theoretical knowledge is validated by conventional peer community of scientists.	Socially useful knowledge is validated by an extended peer community in such a way that personal experience, citizens' concerns and scientific expertise come together to provide a more holistic perspective on science policy issues.
Values are irrelevant (hidden).	Values are central (explicit).
Knowledge alienates passive citizens. Scientific expertise and expert decision-making.	Knowledge empowers critical and active citizens as agents of sustainable development.
Normal science and technocracy remain legitimate so long as environmental and social problems can be solved or ameliorated to the satisfaction of a 'distracted' electorate.	New technologies and the associated risks intensify a legitimation crisis and prompt new forms of public consultation, participation and governance that encourage post-normal science.

Figure 12.1   Normal and post-normal science
Sources: Based on Eden, 1998; Sardar, 2000.

12.3 Four environmental doctrines

Further insight into the roles of normal and post-normal science in environmental politics and policy can be gained by considering MacNaghten and Urry's (1998) classification of environmental doctrines or discourses (Figure 12.2). The dominant doctrine of environmental realism draws on normal science and acts as ideology by masking the social structures and processes that reproduce instrumental or egocentric relations with the rest of nature without recourse to genuine reason or democratic political debate. It separates nature and the environment from society; adopts an acultural, standardized modernist

Naïve environmental realism	The environment is a 'real entity' which separately from social practices and human experiences has the power to produce unambiguous, observable and rectifiable outcomes. Modern rational science provides understanding of that environment and predicts and assesses measures necessary to rectify environmental problems.
Environmental idealism	Analysing nature and the environment entails identifying, critiquing and realizing various 'values' that people hold that relate to the character, sense and quality of nature. Such environmental ethics are treated as stable and consistent and are abstracted from the social practices and groups that may or may not articulate or adopt such ethics. Solving environmental problems entails adopting an appropriate environmental ethic.
Environmental instrumentalism	It is necessary to consider human motivation in order to understand the behaviour of individuals and groups towards nature and the environment. Calculations of individual and collective interests, such as those provided by cost–benefit analysis, suggest ways of solving environmental problems.
Critical environmental realism	Social practices produce, reproduce and transform different natures, environments, environmental issues and values. Such practices mediate the impact of biophysical processes and limits on society and shape people's cognitive and aesthetic responses to what are constructed as environmental issues and representations of nature. They embody their own forms of knowledge and understanding and undermine a simple demarcation between objective science and lay knowledge. Solving environmental problems involves democratizing social practices, knowledge construction and application.

Figure 12.2   Four environmental doctrines
Source: Based on MacNaghten and Urry, 1998.

conception of environmental problems that resonates with bureaucratic governance by state and superstate systems; and thereby facilitates the colonization of nature through technology via such channels as environmental economics, environmental management and green consumerism.

Such realism (labelled 'naïve' in Figure 12.2 to distinguish it from critical environmental realism) also encourages and gains support from the kinds of idealist and behaviouralist social science that underpin the doctrines of environmental idealism and environmental instrumentalism. Like positivist natural science, these fail to acknowledge that environmental issues are fundamentally issues of social relations; that complex social and cultural changes are reconstructing human and non-human nature; and that any realistic science, politics and policy should acknowledge and build upon people's lay and tacit knowledge. Naïve environmental realism, environmental idealism and environmental instrumentalism all misrepresent human engagement with the environment and the complex processes whereby people produce and consume environmental knowledge as they live their everyday lives in diverse social and geographical contexts.

Critical realism and post-normal science attend to these processes and break down the barriers between science and lay knowledge by promoting reflection and action (praxis, p. 39) that allows people to construct socially useful knowledge or **citizens' science** (Irwin, 1995). The global coalition of farmers, consumers, environmentalists and academics against GM foods has done this as it has challenged the power of corporations and governments, and suggested new ways of using biotechnology and information technology to reduce people's alienation from science, politics and policy-making.

12.4 Biotechnology and sustainable development _____

Biotechnology is a broad term used to describe the production of innovative products and services by the exploitation of living organisms and their components. It is based on the manipulation of the genetic structure of cells, draws on advances in biochemistry and molecular biology, and is widely applied in the production of food, chemical feedstocks, alternative energy sources, waste recycling, pollution control and medical and veterinary care. While biotechnology describes a number of novel techniques it has become synonymous in the public mind with genetic engineering or the transfer of a specific piece of genetic information from one organism to another. A genetically modified organism (GMO) is one in which the genetic material has been deliberately altered by humans through genetic engineering rather than through traditional breeding and cross-breeding programmes.

In the 1990s agricultural research institutes and agrochemical corporations invested heavily in biotechnology, suggesting that it was the key to more sustainable ways of increasing the carrying capacity of the land. Farming with GMOs that could fix their own nitrogen, resist pests without pesticides, and grow in difficult conditions, would entail fewer environmental costs. Biotechnology offered new ways of removing food production from the land, providing new uses for some agricultural commodities, bringing marginal into production, overcoming seasonal and time constraints, and so feeding a rising global population. Much research and development focused on engineering plants to produce their own insecticides or to be herbicide-tolerant. By 2000 around one million hectares of land was planted with GM crops worldwide (United States 72 per cent, Argentine 16 per cent, Canada 10 per cent of total), with soya, maize and cotton being the most common of the 12 crops then grown commercially. United States companies had patented the genetic codes of such plants as basmati rice, turmeric, black pepper and cotton, and concerns over biopiracy (Shiva, 1998) are part of the case against GMOs (Figures 12.3 and 12.4). The United States agrochemical giant Monsanto had sought to expand its operations in Europe, but met with unexpected opposition (Box 12.1).

Environmental concerns

1. Most agricultural crops have toxic ancestors and introduced genes could switch back on ancestoral genes making agricultural crops toxic.
2. Genes inserted into GMOs will spread to other non-target organisms with unknown and unpredictable consequences. The extent of such transfer is not clear. Some GM crops are capable of forming resistant populations on waste ground and acting as a continuing source of genes that may transfer to wild relatives through hybridization. This could mean genetic contamination of the gene pool of native plants and animals and a reduction in biological diversity.
3. We do not know enough about ecological interactions to be able to predict accurately the long-term consequences of the introduction of genes into the environment. Small genetic changes can cause large ecological changes. Any changes may not be noticed easily or quickly. GMOs should be bred and monitored through many generations before release, but commercial considerations do not allow this.
4. Development of herbicide-resistant plants could cause changes in the patterns of herbicide use in agriculture in ways that will be more environmentally damaging than existing systems.
5. It is difficult to predict what will turn a plant into a weed. Once an organism becomes a pest it can be difficult to eradicate.
6. A gene does not necessarily control a single trait. A gene may control several different traits in a plant and the placement of genes is a very imprecise science in many cases.
7. There are problems of scaling from field trial to commercial release. A substantial difference exists between a closely monitored experimental plot and a large-scale commercial release. A gene which is safe in one country and one soil type, may behave differently under changed conditions.
8. The majority of the new GM crops require high-quality soils, high investment in machinery and increased use of chemicals. They do not solve the food needs of the world's poorest people.
9. The development of GM crops ignores the social conditions of most of the world's farmers. GMOs encourage continuous cropping and thus discourage rotations, polycultures and the conservation of biodiversity. They work against existing trends towards organic agriculture, extensification and integrated pest-management in the North. Governments in the South, notably India, have attempted to ban GMOs because they wish to protect their traditional agriculture and the food supply of the poor.

Socioeconomic concerns

1. Genetic engineering is leading to the patenting of life forms, genetic information and indigenous knowledge of local ecology. Such commodification or privatization of nature and knowledge is morally wrong. The development of agricultural and food technology should not be left to the free market.
2. Corporations are concentrating research and development on the most profitable elements of biotechnology rather than the applications that best promote sustainable development.
3. While biotechnology has the potential to increase the variety of commercial plants, the trend is towards greater dependence on fewer plants. The control of the global food economy by fewer large corporations is leading to more genetic uniformity in rural landscapes.
4. Competition to gain markets and hence profits is resulting in companies releasing GM crops without adequate consideration of the long-term impacts on people or the ecosystem. Government regulators may not be sufficiently independent and may find it difficult to delay or prevent such releases.
5. Without adequate labelling consumers have no choice as to whether they eat food derived from GMOs.
6. There is no conclusive evidence that GMOs are superior to conventional crops. They may divert resources from exploring more appropriate, low-technology alternatives to intensive agriculture that are more sustainable.
7. Using GMOs to increase agricultural productivity in the North may lead to reduced imports of food from the South. Farmers in the South may then turn to more environmentally damaging alternatives with adverse effects on biodiversity.

Figure 12.3   The case against GMOs

Figure 12.4 An NGO campaigns against patenting on crops
Source: Action Aid.

Box 12.1 Monsanto pays GM price

> Our confidence in this technology and our enthusiasm for it has, I think, been widely seen – and understandably so – as condescension or indeed arrogance (Robert Shapiro, head of Monsanto, September 1999, quoted in Borger, 1999).

Monsanto is a transnational corporation based in St Louis, Missouri, which markets a wide range of products from its chemical and life sciences divisions. In 1996 it employed 28,000 people in 100 countries, and was busy developing such genetically engineered products as Bollgard cotton, New Leaf potatoes and Yieldgard corn, all designed to be resistant to specific insects. It was also buying up other biotechnology and seed companies in what it described as 'a consolidation of the entire food-chain'. It spent around $500 million to develop Roundup Ready soya beans which contain foreign genes from a virus,

(continued)

(continued)

Figure 12.5   Action against field trials of GM crops in the UK
Source: Still Pictures.

a petunia and a bacterium. The beans are designed to be resistant to Monsanto's Roundup, the world's best-selling herbicide, which in the mid 1990s contributed 40 per cent of the corporation's operating profit. Monsanto had a poor record on environmental pollution in the United States, yet was promoting itself as part of the answer to the world's food and environmental problems. Despite protests from environmentalists, the United States Food and Drug Administration approved the marketing of GM crops and by 1998 half the soya and one-third of the corn grown in the United States was genetically modified. GM crops were in just about every processed food, but without relevant labelling – most consumers were oblivious of this fact.

After the EU first approved the import of GM foods in 1996 there were widespread protests by environmentalists and consumers. The popular media printed scare stories about 'Frankenstein Foods' and consumer boycotts led to food manufacturers and supermarkets adopting voluntary labelling schemes or 'non-GM' policies. There were direct action protests against trial fields of GM crops (Figure 12.5) and calls by health professionals and environment and development NGOs for a moratorium on their commercial development to allow more time for research. The level of resistance in Europe surprised Monsanto and in the summer of 1998 it ran a £1 million advertising campaign designed to transform public opinion in Britain. The Advertising Standards Authority subsequently upheld four complaints against Monsanto for claims that this contained.

(continued)

(continued)

Monsanto's fortunes suffered a severe blow after a meeting of EU environment ministers in Brussels in June 1999. In France, Greenpeace persuaded President Chirac and Premier Jospin's advisers to meet separately with Jeremy Rifkin, a prominent United States environmental activist and scourge of Monsanto. He suggested to both president and premier that by supporting a moratorium on new GM applications they would be seen to be caring for people's health and the environment. They should not repeat the mistake of the Blair government in Britain that had seriously misjudged the GM issue and allowed the Conservative opposition to score a rare political coup. France went to Brussels proposing a moratorium, and although this was defeated (largely due to Britain's opposition) ministers did approve compulsory labelling of products with GM ingredients. France, Italy, Denmark, Greece and Luxembourg also declared they would block the further issue of licences until new regulations were in place, in effect a moratorium for at least two years.

The European blockade had immediate effects elsewhere in the world. Brewers and food manufacturers declared that they would not buy GM corn, non-GM crops began to command a price premium, and United States farmers began to cut their GM acreages. Monsanto stock fell from a high of \$50 in May to \$38 in October 1999, and by the end of the year the corporation announced that it would 'spin off' up to 20 per cent of its GM business in a merger with Pharmacia & Upjohn, the drugs company. At that time industry observers believed the GM food's time would come again after a pause for more rigorous testing, and Monsanto was seeking advice from its critics on how best to restore its credibility.

> This was a cultural victory for Europe, where food, cuisine and culture are intertwined, over America, where food is just another commodity. It was seen as an attack on cultural diversity. Cultural and biodiversity is converging into one issue. Food is the last thing people feel they can control (Jeremy Rifkin, quoted in Borger, 1999).

Sources: Borger, 1999; Rhodes, 1999; Toub, 1997.

While private corporations and their allies in government promote a hard path for agriculture with GM crops, others promote a soft path (Figure 12.6). The hard path, based on modern science, is privately financed, centralized and corporately controlled. It leads to ecological monocultures and claims to involve less labour and pollution. The soft path, based on postmodern science, is on the other hand, publicly owned, decentralized and democratically accountable. It is more culturally acceptable because it draws on traditional and new knowledge and technology, and more desirable since it sustains ecology, employment and culture. The vision of soft paths to an agricultural future with biotechnology has inspired India's public agricultural research institutes. Professor Swaminathan at the Centre for Research on Sustainable Agriculture in Madras believes that such advances as salt-tolerant rice varieties, made freely

Hard path	Soft path
Nature as external to society and to be 'tamed', 'mastered' and 'controlled'.	Nature as a seamless web of symbiotic relationships and mutual dependencies that includes society or human nature.
The world is seen in reductionist terms and scientists regard themselves as grand engineers, continually editing, recombining and reprogramming the genetic components of life to create more compliant, efficient and useful organisms that can be put to the service of humankind.	The world is seen in dialectical and systemic terms with the earth and its living things constituting a single (differentiated) organism – the biosphere. Scientists and others should engage in subtle forms of manipulation that enhance rather than sever existing relationships.
Molecular biologists insert alien genes into the biological code of food crops to make them more resistant to herbicides, pests, bacteria and fungi. They envision these engineered hybrids living in a kind of genetic isolation, walled off from the larger biotic community, and ignore the environmentalist's fears of genetic pollution.	Ecologists use the new genomic information to help them understand how environmental factors affect genetic mutations in plants. They use the new scientific knowledge to improve classical sustainable farming methods, such as breeding, pest-management, crop rotation.
Uses the new genetic science to engineer changes in the very blueprint of species.	Uses the same genetic science to create more integrative and sustainable relationships between existing species and their environments.
Privately financed, centralized, corporate control. Establishes ecological monocultures and erodes biodiversity and human diversity.	State-financed, decentralized, community control. Promotes biodiversity and human diversity.
Promotes academic knowledge over local knowledge.	Values local knowledge.

Figure 12.6 Two paths to the future with biotechnology
Source: Based on Rifkin, 1999a and 1999b.

available to farmers, are an important aid to greater self-reliance and rural sustainability (Vidal, 1999).

The women members of the Deccan Development Society spread over 40 villages 100 miles from Bangalore, have demonstrated the advantages of the soft path while rejecting any use of GM crops. By combining traditional farming with the best new research, and applying such techniques as permaculture, composting, intercropping, water-conservation and manuring, they have within a decade restored degraded lands, doubled the number of crops they grow, and increased yields by 50 per cent or more. Many in the villages are still landless and unemployed, but ecologically based agriculture has brought more self-reliance and strengthened opposition to GM technology. As British farmers face an agricultural crisis of overproduction and falling real incomes, environmentalists and others are urging them to consider a similar soft or organic approach. Governments in India and Britain should undertake a broad analysis of the benefits and risks associated with GM technologies and articulate a vision of sustainable rural development that balances attention to employment, environmental protection, nature conservation, healthy communities and the production of healthy food. An examination of the Government's handling of the GM food issue in Britain, strongly based on the work of the Centre for the Study of Environmental Change (Scott, Berkhout and Scoones, 1999), suggests that failure to promote such a vision contributed to the public's lack of confidence and trust.

12.5 Science and uncertainty

While environmental realism suggests that science can provide objective and neutral advice on such issues as GM food, the way that scientific knowledge is used is heavily influenced by the way official advisory systems operate. Those who advise governments may adopt too narrow a view of the sorts of knowledge required to make judgements (excluding ecologists from advisory committees for example) and may falsely assume that the safety and acceptability of commercial GM agriculture can be settled by science alone. Wynne (1999) and others challenge this assumption, suggesting that standard risk-assessments are unable fully to characterize the fundamental risks and uncertainties associated with the potential impacts of such new technology. This is because knowledge generated by science in controlled experimental conditions is not necessarily valid in commercial field situations. What is more, not much is yet known about many aspects of GM agriculture, and there are all sorts of things we don't know that we don't know. Scientific understanding of GMOs is inherently provisional, and since probabilities and outcomes cannot be well defined or quantified, it may be more appropriate to speak of **uncertainty or ignorance** rather than risk. Scientific advisers should be open about the limits of understanding and politicians should not represent a lack of evidence of risk as evidence of no risk.

Clearly the sources of some of the uncertainty surrounding GMOs are economic and social, and the subject of research in the social rather than the natural sciences (Powell *et al.*, 1997). This suggests that scientific judgements on risk and uncertainty are underpinned and framed by unavoidably subjective assumptions about the nature, magnitude and relative importance of these uncertainties. These '**framing assumptions**' can have an overwhelming effect on risk-assessments, explaining why different assessments of the same issue, all based on 'sound science', can lead to widely varying results. The adoption of any particular set of framing assumptions cannot be justified in terms of 'science', but must be assessed in terms of such factors as the legitimacy of the institution making the justification; the degree of democratic accountability to which the institution is subjected; and the ethical acceptability of the assumptions adopted.

12.6 Public perception of risk

In the light of previous reassurances about the risks associated with such things as lead, asbestos, nuclear power and beef, the British public now mistrusts statements on risk made by of scientific experts, officials and politicians. They in turn express frustration at the public's 'inaccurate and emotional' reactions to such issues as GM food, accusing campaign groups of spreading alarm through media pronouncements based on half-truths or speculation. Research carried out by the Global Environmental Change Programme (Grove-White *et al.*, 1997) suggests that public understanding of risk issues is very much better developed than these criticisms assume. People repeatedly mentioned the BSE crisis to support their unease about biotechnology, suggesting it posed the same type of risks in terms of unnaturalness; the failure of institutions to prevent risks; the long-term character of associated risks; and our ability to avoid them. They may not know the scientific and technical detail, but they have a sharp awareness of the broad issues involved and ways of forming their own judgements. They are particularly mistrustful of the scientific approach to uncertainty that may lead to 'surprises' in the future.

There is nothing irrational about such public perceptions of risk, for people realize that the scientific evidence may be incomplete; the scientists may have vested interests in selecting particular bits of evidence; or there may be more than one interpretation of the facts. While most people are open-minded and hold conflicting and varied opinions about different uses of GM technology, experience has taught them that it is rational to harbour doubts about such technology. When unable to assess the risks themselves, they tend to judge those who create the risks or regulate them. In the late 1990s many suspected, with some justification, that the British government's open support for biotechnology meant that it had sided with the industry, and that its advisers and regulatory agencies were not independent. The case of Monsanto (Box 12.1) suggests that the company and British government underestimated public mistrust of GM foods and that **trust** may be more important than knowledge in deciding how people judge new technologies. It is the scope, powers and efficacy of the regulatory system that largely determines such trust.

12.7 The regulation of GM foods in Britain and Europe _____

There is a comprehensive EU-wide regulatory framework covering all stages of work with GMOs that is designed to protect human health and the environment (EU Directive 90/220). All GMOs are subject to a safety evaluation before consent for release into the environment is agreed, and in Britain the Secretary of State for the Environment issues such consents. S/he is advised on the release and marketing of GMOs by the statutory Advisory Committee on Releases into the Environment and by other EU and national bodies, but because they have assessed products on a 'one at a time' basis, using limited scientific evidence framed in particular ways, they have not provided a comprehensive, scientifically robust, foundation for regulatory decision-making. Of particular concern in the 1990s were their narrow remits that precluded such issues as the need for GM foods and the social benefits envisaged from

GM strategies; the potential for indirect ecological and health effects; and the wider consequences for agriculture and the countryside.

In July 1999 the British government announced two new strategic committees on GM food to consider cross-cutting issues, but although the Human Genetics Commission and the Agriculture and Environment Biotechnology Commission were to have a wider membership, shortcomings about assumptions and working practices remained. The first international treaty regulating trade in GM products was agreed in 2000, allowing nations to ban imports on social, environmental or health grounds, but it could take years for the treaty to be translated into national laws. Meanwhile Prime Minister Blair continued to support President Clinton by calling for the introduction of GM foods guided by the 'best science', but other European leaders reminded him of the importance of abiding by the precautionary principle.

12.8 Democratizing environmental knowledge

On the issue of GM food it is clear that dominant forms of science and decision-making have failed to foster sustainable development and prompted a crisis of trust or confidence in large sections of the public. Researchers in the Global Environmental Change Programme suggest that the government needs to:

- Restore its neutrality in the eyes of the public.
- Address gaps in the advisory framework (as outlined above).
- Broaden the intellectual 'problem framings' that shape current scientific advice, in order to address the weaknesses of narrow 'sound science' judgements on risk.
- Be more attentive to social factors, rather than dismissing public unease as a matter of irrationality and lack of understanding.
- Explicitly take account of the time-scale issue and the related questions of how any potential future costs and liabilities might be handled.
- Systematically analyse the effects that human judgements, such as those shaping scientific assessments on risks, have on the results of these assessments.

Rather than simply widening the membership of advisory committees, the researchers suggest that there should be a comprehensive audit of the diversity of interests, assumptions and values that characterize the different interested and affected parties. **Deliberative and inclusive processes** (Figure 12.7) allow such information to be gathered, interpreted and verified, and these are well established in other countries as part of the regulatory appraisal of technological risks. They offer ways of listening to lay knowledge, generating citizens' science, bringing to light the different ways of interpreting scientific advice in the regulation of risks and can complement the work of advisory committees.

There is little point in promoting the development of citizens' science in such ways unless the government has clear criteria and mechanisms for listening to a wider range of advice. Governments will need to broaden the scope of

Focus groups typically consist of 6–10 individuals. A trained facilitator takes the group through a number of set questions around an issue while maintaining informality. The group discusses these but there is no pressure to make decisions or recommend actions. The purpose is to record and analyse the discussions to reveal people's main concerns.

Citizens juries involve 12–25 participants in an in-depth evaluation of alternative policy options. Chosen to represent the local community, the citizens hear evidence to support different options from a wide range of witnesses (local authority officers, scientists, pressure groups, business managers, etc.). They can challenge witnesses and formulate new or revised options, but must eventually reach a decision.

Other techniques include **consensus conferences, visioning conferences, stakeholder decision analysis, young people's hearings** and **deliberative polling**. All seek to create small public spaces (public spheres) where people can discuss their beliefs and values with each other and with decision-makers.

Figure 12.7   Deliberative and inclusive procedures

regulatory appraisal; conduct assessments on a comprehensive rather than 'one at a time' basis; maintain a culture of humility and pluralism in the face of uncertainty and ignorance; provide for open-ended interaction between scientific monitoring and analysis, and inclusive deliberation; and uphold the primacy of institutional legitimacy and political accountability in the final justification or regulatory decisions. The UK government should meet these objectives through coordinated reform of its policies on agriculture, food, the countryside and sustainable development that allows for the possibility of deciding not to go ahead with GM food technology.

12.9 Future governance of new technologies

The British government's experience of GM risk-management points to the advantages of adopting a **precautionary approach** whereby governments act with a duty of care when unambiguous scientific proof of cause and effect is not available. This requires early action when faced with evidence of the likely costs of delay; protection of critical natural capital and services threatened with irreversible damage; openness to calls for a change of policy from what-ever source; freedom of information; and the promotion of discursive demo-cracy to dispel public unease. As we have seen, 'sound science' is a necessary but not sufficient basis for decisions, and governments should experiment with new policy- and decision-making techniques and processes that accom-modate other kinds of knowledge. These should seek to protect those least able to protect themselves and should acknowledge the inevitably provisional nature of decisions on issues pervaded by ignorance and uncertainty. Research suggests that the opinions and judgements of increasingly reflexive individuals are influenced by diverse social networks and the creation of trust in new technologies and regulatory systems may require different messages aimed at different audiences. **Collaborative research** between private companies, pressure groups and government is revealing the values and assumptions that citizens consider should be at the heart of the regulatory process, and reform to reflect these is the key to restoring the legitimacy of government and companies such as Monsanto.

Reforming the governance of new technologies, so that it is guided by citizens' science, requires politicians to share power at a time when most remain sceptical about participatory or community democracy. It requires regulatory agencies, companies and government departments to become more open in providing information and allowing access to decision-making, and all parties in disputes to accept greater responsibility for working towards consensus. The new information and communication technologies can encourage such discursive democracy.

12.10 The Internet and cyber–democracy

Chapter 9 suggested that time–space compression, enabled by the communications revolution, is fostering a global civil society, or anti-corporate movement, that resists the deregulation of the global economy. Operation Cremate Monsanto in India (Box 12.2) provides just one example of how workers,

---

Box 12.2 Operation Cremate Monsanto and cyber–democracy

Plant seed is economically and culturally significant for India's 500 million farmers. Eighty per cent of the seed from agricultural crops is collected and replanted, serving as a sustainable source of livelihood that is celebrated in religious and other festivals. The farmer's right to save, reuse and share seed is closely guarded against commercial interests that seek to gain control of seed supply.

Cotton is politically significant in India. Freedom from cotton colonization was a central argument for independence 50 years ago when the wearing of hand-woven cotton cloth symbolized opposition to British rule. Gandhi envisioned a decentralized self-sufficient India, but the country remains highly dependent on exports of cotton. India is the world's largest producer, earning almost one-third of its foreign exchange from cotton.

In the late 1980s and early 1990s the IMF and World Bank persuaded India to open its heavily protected economy to world trade. In return for loans it was made to dismantle its immense state-owned seed supply system, run down subsidies and public agriculture institutions, and give incentives for the growing of cash crops to earn foreign exchange. Several million farmers were persuaded to turn from traditional crops to cotton and use high-yielding varieties that required expensive pesticides and other inputs (Figure 12.8). By 1999 up to 60 per cent of the cultivable land in Andrah Pradesh was planted with cotton. When the crop failed that year, hundred of farmers who were in debt to moneylenders committed suicide.

(continued)

(continued)

Figure 12.8 Farmer with Roundup herbicide
Source: Adrian Arbib.

Monsanto has been making and selling herbicides in India since the 1960s. In the 1990s it began to promote GM agriculture as sustainable development that would reduce costs, increase yields, reduce pesticide use and bring 'Food, Health, Hope'. To this end it purchased India's most advanced genetic engineering research centre for $20 million and spent $4 million buying up seed companies. At the time when India's cotton farmers were experiencing problems with existing high-yielding varieties, Monsanto promoted its Bollgard

(continued)

(continued)

variety. Advertising and sports sponsorship were used to draw attention to its claimed advantages of bollworm resistance, higher yields and reduced pesticide use, but farmers and environmentalists became increasingly suspicious of Monsanto's claims. They learnt that Monsanto had acquired the United States seed giant Delta and Pine that dominated the United States cotton market and owned the United States patent on the 'terminator gene' that allowed plants to be genetically engineered to produce sterile seed. While Monsanto denied that it had any plans to incorporate 'terminator technology' into cotton, the fact that it now owned the patent was sufficient to alarm a coalition of nationalists, Gandhians and Marxists and prompt direct action against trial sites that they claimed Monsanto had planted without correct permissions or safety procedures.

In November 1998 Karnataka state farmers, led by Professor Najundasmamy, a self-styled Gandhian, launched its Cremate Monsanto campaign. As well as burning Monsanto's trial fields of Bollgard cotton, farmers occupied and pulled down a Cargill seed factory, and organized a 'laugh-in' that led to the downfall of the state government. Najundasmamy promotes the decentralization of farming and politics, has organized rallies against such organizations as the World Trade Organization and Kentucky Fried Chicken, and uses the new communication technologies to link his campaigns and ideas with others across the world. During the Cremate Monsanto campaign he used a Website (Peoples' Global Action http:// www.agp.org) to explain how neoliberal globalization undermines environmental and food security. Such analysis is echoed on a large number of related Websites and the associated e-mail lists and newsgroups sustain a virtual community that is gradually acquiring communication, media and lobbying skills to match those of the biotechnology corporations.

Source: Vidal, 1999.

consumers and environment and development NGOs are using new technologies to share information and strategy, critique the environmental discourses promoted by companies and governments, and create global coalitions for change. While access to the Internet is still largely confined to more affluent citizens in the North, wider access to corporate (e.g. www.monsanto.com/), government (e.g. www.maff.gov.uk/), NGO and media sites (e.g. www.oneworld.org/, www.connectotel.com/gmfood, www.gmissues.org/ frames.htm, www.newsunlimited.co.uk/gmdebate) should serve to widen and deepen democracy and generate more of the kind of socially useful environmental knowledge or citizens' science advocated in this chapter. The new technologies encourage this by allowing the individualization of information consumption and erosion of the power of the mass media. The rise of **cyber-democracy** is an example of what Lash and Urry (1994) describe as culture coming to the rescue of nature. Virtual communities in cyberspace contest dominant forms of environmental knowledge and risk-management in order to defend or restore valued forms of identity, community and nature. The regulation of cyberspace and its domination by commercial interests are real possibilities, but for the time being the Internet represents a powerful tool for fostering the kind of global citizenship that is the key to sustainable development.

12.11 Further reading

The booklet *Food for our Future* (Food and Drink Federation, 1999) seeks to provide consumers with clear and comprehensible information about biotechnology. Does it succeed in its aims?

Porritt (2000) demolishes notions of an objective, value-free environmental science that can guide public policy, and suggests that the growing commercialization of science may render it unfit to rise to the challenge of sustainable development. Read the book and debate his arguments.

With reference to Naughton (1999) and other sources, decide on the main contradictions surrounding the use of the Internet as a tool for communicating and campaigning on environmental issues.

12.11.1 Websites

Global Environmental Change Programme: http://www.susx.ac.uk/Units/gec/
Centre for Science and Environment, India: http://www.cseindia.org

Chapter 13

Towards sustainable development

13.1 Introduction

Having arrived at the penultimate chapter of this book you should have a clearer understanding of how the structures and processes shaping past and present societies cause them to develop in more or less sustainable ways. Previous chapters have suggested some sustainable alternatives to current realities, but this chapter hopefully offers more coherent and considered grounds for cautious optimism. It begins by considering the contradictions of our current environmental predicament and then links the prospects for future sustainability to new forms of global governance and citizenship. Such governance would restructure the rights and responsibilities of private corporations, governments, groups within civil society and individual citizens, and consideration of ecological modernization, together with the UK government's strategy for sustainable development, suggests that it should go beyond reformism. Only by radically democratizing economic, political and cultural structures of power, and revitalizing civil society, can we move to sustainable societies in which people are content to live within ecological limits because they recognize the virtues of simplicity and frugality. Geographers have roles to play in guiding such a transition to sustainability, and the chapter concludes by considering some of the contributions they are making and challenging readers to act on some of the critical ideas presented in this book.

13.2 A world of contradictions

Chapter 1 provided an overview of the contradictions surrounding the environment and development in the contemporary world (Box 1.2, p. 14). You are now better able to link these contradictions to the dynamics of global capitalism and the neoliberalism that has accompanied its recent restructuring. The drive to accumulate capital continues to destroy sustainable ecological and social relations and erode those safeguards established in the name of socialism to protect human, ecological and social capital. National forms of social democracy and state collectivism have broken down under the pressure of globalization since economic and political elites are less able and prepared to care for the poor and the environment within national borders. As capitalists avoid paying taxes, creating jobs, reinvesting surplus or reproducing the conditions of production, and political leaders link nations' destinies to global markets claiming there is no alternative, the promise of development, the prospect of social consensus and the legitimacy of the nation-state all fade.

The reality of today's world is not a universal market but an anarchy of sovereign states, rival capitalisms and stateless zones. The modern state has not taken root or has collapsed in much of the world and here a lack of effective governance offers little prospect of reconciling the imperatives of global markets with the needs of social cohesion and environmental conservation (Gray, 1998). Regional and international bodies of global governance such as the European Union, G8, OECD, IMF, World Bank, World Trade Organization and UN lack a democratic mandate, are strongly influenced by powerful minorities and generally act in the interests of capital rather than promoting social justice or sustainability. Anarchic market forces and shrinking resources drag nation-states into ever more dangerous rivalries, yet the regulatory agenda is weak and in urgent need of strengthening.

Chapter 9 introduced regulation theory as an approach to the changing relationships between the process of capital accumulation and the ensemble of institutional forms and practices which together guide and stabilize the process and create a temporary resolution of its crisis tendencies. It explained that **modes of social regulation** take a variety of forms, operate at different levels of abstraction, may be outcomes or causes of economic change and serve to institutionalize such struggles as those between citizens' and workers' movements and the state. The establishment and growth of a new wave of (post-Fordist, postmodern) accumulation in the global economy requires and shapes new modes of regulation with the dominant neoliberal response serving to accentuate social and environmental problems. Sustainable development is an emergent alternative that claims to resolve the impasse between economic development, environmental conservation and social inclusion, and takes such forms as the patchwork of international environmental agreements, corporate environmentalism, green consumerism and the incorporation of sustainable development into local and national economic policy (Gibbs, 1996).

As an emergent mode of regulation, sustainable development is being shaped by conflict and struggle. The shift to more flexible forms of production and consumption, based on new information, communication and biotechnologies, will not necessarily bring greater sustainability, as much utopian literature assumes. The new economies have the potential to be less polluting, less demanding of energy and materials, and more decentralized and democratic, but whether or not they realize this potential and liberate human and non-human nature from current forms of domination depends on the outcome of struggles by workers' and citizens' movements around the world. Only their coordinated efforts can build those forms of regulation that constitute a **cosmopolitan democracy** and will enable strong sustainability (p. 235) to take root and grow (Beck, 1998; Graham, 2000).

13.3 Cosmopolitan democracy

In urging new forms of democracy, citizenship and governance, radical environmentalism recognizes that the construction of nature and sustainability should become matters of explicit and discursively justified communal choice. As we saw in Chapters 1 and 2, sustainability is essentially an issue of social relations, their impact on technology and ecological relations, and on our physical, mental and social health. Democratic social relations allow open and equal discussion and decision-making in which theory can be combined with practice and all voices, including those of future generations and the rest of sentient nature, can be heard (Dryzek, 1996). Such decision-making allows the rational consideration of what is technically possible, culturally appropriate, and morally and politically right, and is a legitimate means of translating agreed ethical principles into the rights, responsibilities and legislation that form the basis of citizenship. It is the vehicle of revolutionary praxis (p. 39) that generates citizens' science and guides much community development (Rahman, 1993). The deliberative and inclusive procedures (Figure 12.7, p. 225) considered in the last chapter suggest that states are now prepared to experiment with more participatory forms of democracy and such procedures form a central element of much planning for sustainability.

David Held's outline of cosmopolitan democracy and a new international order (Held, 1995) helps us to envision the changes that will be necessary before democratic social relations and discursive decision-making are the norm at all levels and across all sites of power. Following the tenets of dialectical materialism and critical realism, he reminds us that the global order is constituted by multiple, overlapping, dynamic sets of social relations or networks of power. These networks contain seven sites of power that shape people's capacities and life-chances, the kinds of technology and discourse that mediates their relations with one another and the rest of nature, their rights and duties, and hence their status as citizens.

The seven sites of power (and the aspects of people's lives that they condition) are:

- the body (physical and psychological well-being);
- social welfare (opportunities to become an active member of the community);
- culture (cultural identity);
- civil society (opportunities to join civic associations);
- the economy (capacity to influence the economic agenda);
- coercive relations and organized violence (ability to act without fear of physical force and violence);
- regulatory and legal relations (ability to participate in political debate and electoral politics).

In the modern period **citizenship** (the framework of complex interlocking relations which exist between rights and duties in any legal and moral system) expanded to embrace civil, political and social citizenship. Citizens have acquired

- Global governance should be based on **the principle of autonomy**. All the world's people should enjoy equal rights, and accordingly equal obligations, in the specification of the political framework which generates and limits the opportunities available to them. They should be free and equal in the determination of the conditions of their own lives, so long as they do not deploy this framework to negate the rights of others. People should be self-determining and democratic government should be limited government. It should allow 'the people' to determine the conditions of their own existence while limiting 'the people's' power through a regulatory structure that is both constraining and enabling.
- Enactment of the principle of autonomy requires an expanding framework of legal principles, institutions and procedures, to extend and deepen democratic accountability at all levels from the local to the global. These can provide and enforce rights and responsibilities that cut across networks of power and provide the foundation for new forms of **global democracy, governance and citizenship**. Laws would delimit the form and scope of individual and collective action within the organizations and associations of the state, economy and civil society, creating minimum standards for the treatment of all, and ensuring the effective coordination of social development in the common interest.
- Global democracy could reshape and redistribute political powers. It could **recast territorial boundaries of accountability** so that issues and agents which currently escape the control of nation-states could be brought under democratic control. It could **reform regional and global regulatory and functional agencies** to give them a more coherent and powerful role in realizing sustainable development. It could also **ensure that key groups, associations and organizations, from within the economy and civil society, become part of the democratic process**, at all levels from the local to the global. Such changes will require an expansion of the influence of regional and international courts to monitor compliance with an expanded framework of legal principles.
- Global democracy could ensure that the exploitation of natural resources and services and the production and distribution of economic wealth is **constrained within ecological limits** and takes place according to **principles of social justice and sustainability**. It could **use the principle of non-coercive relations to govern the settlement of disputes**, using force only as a collective option of last resort in the face of clear attacks on cosmopolitan democratic law.

Figure 13.1   Towards a cosmopolitan world order
Source: Based on Held, 1995.

legal, political and welfare rights, along with corresponding duties, and each phase of development has been associated with particular ideas of justice. The primary container of citizenship has been the nation-state, but the growth of global networks of power, the urgency of global issues such as climate change, and the emergence of local groups, movements and nationalisms from below, now challenge the power and legitimacy of the nation-state and that of the present undemocratic interstate system. Calls for new postmodern systems of global governance and citizenship are intensifying, and Held suggests a model for our consideration (Figure 13.1).

In seeking to embed the **principle of autonomy or self-determination** into all sites of power, at all levels from the local to the global, Held's model seeks to extend further the depth and breadth of citizenship. He wishes us to have rights and responsibilities across all aspects of our lives (all sites of power) and for these to be guaranteed and made real by governments and other

- **Increase public participation** The global financial system is run by bureaucrats, bankers and mainstream economists. Their decisions have a profound impact on the lives of ordinary people who are never consulted. The institutions which determine global trade and investment policy have an obligation to incorporate the opinions and insights of **civil society** into their decision-making.
- **Establish a global financial authority** National governments have lost control of their ability to manage their own economies. The world needs a new international regulatory agency to reduce volatility and inefficiency in global financial markets – a **Global Central Bank**.
- **Stop speculation** Fast-moving, unregulated investment has turned the global economy into a casino where big-money speculators search for instant profits and ignore the consequences. A **speculation tax** that would put people ahead of profits is urgently needed.
- **Control capital** New global investment rules give carte blanche to investors while demanding nothing in return. An **alternative investment code** is needed for democratic control of capital and to stimulate investment that benefits local communities.
- **Honour the Earth Global environmental standards** must be set by a new world body under the mandate of the UN – a Global Environmental Organization. These standards must be based on sustainability, equity and justice, and should be embedded in all international trade and investment agreements.

Figure 13.2   Redesigning the global economy
Source: Ellwood, 2000.

institutions at all levels from the local to the global. Cosmopolitan democracy and citizenship allows effective coordination of social development in the common interest, will require the redesign of the global economy along the lines suggested in Figure 13.2, and is likely to lead to the protection, conservation and restoration of biophysical resources and services in the interests of present and future generations. To use one of Held's examples, it would allow factories to be locally monitored and challenged, nationally regulated and supervised, regionally checked for cross-national standards and risks, and globally evaluated in the light of their impact on health, wealth and economic opportunities for others. In such ways an emerging global democracy will embrace an **ecological citizenship** that extends rights to future generations and other members of the biotic community and stems from an enlarged concept of justice (Roche, 1992; Smith, 1998). Some suggest that while the European Union is in need of much reform and democratization, it has the potential to prefigure such a global democracy.

13.4 The discourse of sustainable development

Realizing sustainability through a radical democratization of social relations, new modes of social regulation or new forms of global governance is not a priority for the world's economic and political elites. They frequently use the term 'sustainable development' to legitimate policies and actions that are simply adjustments to the existing order, and set against the above prescription such reformism points to fundamental contradictions in the discourse (Rees, 1985; Redclift, 1987). These can best be understood by contrasting the

Sustainability in the development mode	Sustainability in the growth mode
• Implies a radical democratization of current social relations. New systems of global governance protect the well-being of human and non-human nature while constraining the global economy within ecological limits. • Seeks ecological, economic, social, cultural and personal sustainability by developing ecological, human, social and organizational, and manufactured capital using appropriate technology. Has high regard for critical ecological capital and promotes strong sustainability. • Ensures that social development promotes the continued progressive evolution of human and non-human nature. It facilitates the redistribution of wealth from the rich to the poor and allows local communities a high degree of self-management in realizing their own forms of sustainable livelihood. • Recognizes the value of ecological modernization. • Stresses the role of local community development guided by citizens' science. • Regards ecological limits as enabling. • Emphasizes sufficiency. • Is homocentric or ecocentric. • Allows and promotes the greening of socialism.	• Does not require a radical restructuring of capitalist social relations. Human and non-human nature are viewed instrumentally and sustainability is one goal to be realized along with continued capital accumulation or economic growth. • Seeks ecological and economic sustainability with little attention to social, cultural and personal sustainability. Develops ecological and manufactured capital at the expense of human, social and organizational capital. Is prepared to substitute critical ecological capital for other forms of capital and so promote weak sustainability. . • Includes groups that promote ecological modernization (a shift to more environmentally benign systems of production and consumption) and global welfare through institutional reform and redistribution. • Stresses the role of experts (e.g. ecologists, engineers, economists, planners, lawyers) guided by normal science. • Regards ecological limits as constraining. • Emphasizes efficiency. • Is egocentric or homocentric. • Allows and promotes the greening of capitalism.

Figure 13.3   Sustainable development vs sustainable growth

social construction of sustainability in the development mode with its construction and in the growth mode (Figure 13.3). While the former links sustainability to forms of postmodern socialism, the latter anchors it firmly to postmodern capitalism.

The contrasting meanings outlined in Figure 13.3 permeate, in different ways and to differing extents, official reports, policies and plans for sustainable development. *Our Common Future* (WCED, 1997), the report of the World Commission on Environment and Development (chaired by Gro Harlem Brundtland) updated the case for a global social democracy by suggesting that economic growth, social justice and environmental protection could be combined. Since poverty is a key cause of environmental degradation, environmentally friendly growth (or sustainable development) is necessary to end poverty and protect the environment (Elliott, 1998). Such growth or development should give priority to the needs of the poor and should recognize limits to the environment's capacity to meet present and future needs imposed by prevailing forms of technology and social organization. Such limits are flexible and *Our Common Future* recommends more efficient technologies, new forms of economic accounting, fairer exchanges of goods, services, capital and technology between North and South, reform of global institutions, and greater public participation in development, as means of increasing present and future human welfare while living within ecological limits. The Commissioners expected too much of powerful elites who had largely abandoned social democracy for neoliberalism, but were correct in recognizing that in the final analysis sustainability is about governance. They suggested that it requires:

> . . . a political system that secures effective citizen participation . . . an economic system that is able to generate surpluses and technical knowledge on a self-reliant and sustained basis, a social system that provides for solutions for the tensions arising from disharmonious development, a production system that respects the obligation to preserve the ecological base for development, a technological system that can search continuously for new solutions, an international system that fosters sustainable patterns of trade and finance and an administrative system that is flexible and has the capacity for self-correction (WCED, 1987, p. 65).

The UN Conference on the Environment and Development (the Earth Summit held in Rio de Janeiro in 1992) sought to establish a global partnership and action plan for sustainable development. The largest-ever gathering of world leaders endorsed *Agenda 21* (UNCED, 1992a) (Figure 13.4), issued the *Rio Declaration on Environment and Development* (UNCED, 1992b), and agreed conventions for climate, biodiversity and deserts, and a set of guidelines for forests. A Global Environmental Facility and a Commission on Sustainable Development were also established to manage the funding of sustainable development projects and monitor progress in realizing the objectives set out in Agenda 21. The summit marked a watershed in how the global community tackles environment and development issues, succeeded in raising awareness of the need for sustainable development and emphasized the key role of groups within civil society. It failed to tackle overdevelopment and consumption in the North; to address the regulation of transnational capital; to secure commitments from the North on debt relief, fair trade or the funding of sustainable development; and was too optimistic in expecting national governments to cede power to civil society and supranational bodies. Failure of the North to keep its bargain with the South (to fund sustainable development with aid and technology transfer in return for action on pressing environmental issues) and crisis in the UN limited progress after Rio (Reid, 1995; Dodds, 1997). The Rio agenda has not been properly addressed and integrated across the departments of most national governments, but there has been worthwhile follow-up at national and local levels. Around 60 nation-states established national round tables on sustainable development and thousands of local governments engaged in the Local Agenda 21 (LA21) process.

In the years since the Rio Earth Summit, sustainable development has become the organizing principle or dominant discourse (Figure 1.8, p. 20)

Agenda 21 discusses the substance of what sustainable development should mean, the process through which it can be decided on and achieved, and the management tools needed to achieve it:

Substance	Process	Tools
• reduce use of resources and production of waste, increase resource efficiency, reuse, recycle • conserve fragile ecosystems • social equity (between and within countries and across generations) • quality of life (broader than standard of living) • respect for traditional knowledge, ways of life, diversity	• active planning and management • consultation, participation, empowerment • decisions at most local level possible, local government pivotal • partnerships and collaborations between all sectors	• education, information, awareness-raising • capacity-building, institutional know-how, confidence, experience • regulations and enforcement • market management, taxes, levies, subsidies • public investment

The UK strategy for sustainable development, *A Better Quality of Life* (DTER, 1999), has four aims, seven priorities, and ten guiding principles:

Aims	Priorities for the future	Guiding principles
• social progress that recognizes the needs of everyone • effective protection of the environment • prudent use of natural resources • maintenance of high and stable levels of economic growth and employment	• more investment in people and equipment for a competitive economy • reducing the level of social exclusion • promoting a transport system that provides choice, and also minimizes environmental harm and reduces congestion • improving the larger towns and cities to make them better places to live and work • directing development and promoting agricultural practices to protect and enhance the countryside and wildlife • improving energy efficiency and tackling waste • working with others to achieve sustainable development internationally	• putting people at the centre • taking a long-term perspective • taking account of costs and benefits • creating an open and supportive economic system • combating poverty and social exclusion • respecting environmental limits • the precautionary principle • using scientific knowledge • transparency, information, participation and access to justice • making the polluter pay

Figure 13.4   Agenda 21 and *A Better Quality of Life*
Sources: UNCED, 1992a; DTER, 1999.

of modern environmentalism, allowing capitalism to co-opt many of its environmental critics as modes of regulation adapt to new realities (Jacobs, 1997). Sachs (1999) is scathing in his critique of the dominant discourse suggesting that sustainability in the growth mode (Figure 13.3) facilitates the rise of a global **ecocracy** that seeks to manage pressing problems of risk and security in the interests of powerful elites. This new form of imperialism is necessary to sustain capital's means and conditions of production, and requires global society to be reconstructed as 'one world' in which divisions of race, class and gender dissolve in the face of an imagined common predicament. The discourse of environmental modernization provides the framework in which the new ecocrats respond to the environmental challenges that lie at the heart of the sustainability agenda.

13.5 Environmental modernization

A report to the Club of Rome (Lovins, Lovins and von Wiezacker, 1997) suggests that increasing resource productivity rather than labour productivity is the key to progress in the twenty-first century. Today's technologies allow a fourfold rise in the efficiency of resource use, since an economy that incorporates the best of current ecological designs for products and services can do more with less while reducing pollution and waste (Figure 13.5). Agricultural systems that require minimal inputs, waste-free industrial processes, teleconferencing, paperless offices, photovoltaics, windows that trap heat in buildings, and consumer goods that use much less power, are just some of the innovations that prompt many to suggest that capitalist economies can be restructured along environmentally sound lines. Like other modes of regulation, sustainability in the growth mode (labelled **environmental or ecological modernization**) takes

Figure 13.5 Lack of investment in wind power in the UK means an opportunity for ecological modernization is being lost
Source: Still Pictures.

a variety of forms from the 'real' regulation of laws and concrete structures (e.g. recycling facilities, public transport) through to more intangible elements such as values and norms of behaviour.

Dryzek (1997) suggests that ecological modernization (Figure 1.8, p. 20) works best in those countries, such as Germany and the Netherlands, where corporatism and consensus prevail. Here partnership between government, business, environmentalists and scientists, allows greater economic democracy and a greater likelihood that workers, shareholders and customers will recognize and realize their shared interests in more sustainable forms of production and consumption. Shifting the tax burden from labour to resource use and pollution can encourage such reform, as can appropriate regulation, but supporters of environmental modernization generally prefer to stress incentives and persuasion. By offering increased markets and profits to capital, improved health and safety to workers, greater satisfaction to green consumers, and improved legitimacy and national competitiveness to politicians, such modernization appears reassuring and distracts attention from more difficult choices. The UK Government's Sustainable Development Strategy (DTER, 1999) was much influenced by this discourse, and some urged New Labour's leaders to embrace it more firmly (Box 13.1).

Box 13.1   New Labour and environmental modernization

New Labour came to power in 1997 with a mandate to modernize Britain. It advocated the 'third way' (Giddens, 1998) as a system of national regulation, and some considered it to be Britain's first postmodern government, paying more attention to presentation, style and corporate management than policy, debate and democracy. The environment did not feature in its election manifesto and the new government failed to develop a discourse and policies on the environment to match those on the economy, education, social exclusion or law and order. According to Michael Jacobs (1999) of the Fabian Society, the problem lay in New Labour's suspicious attitude towards environmentalism and its dislike for unrepresentative pressure groups. The majority of its leaders saw environmentalism as anti-aspirational, irrelevant to the voters of middle England, anti-consumption, anti-poor in its advocacy of regressive taxes, and anti-business in seeking environmental taxation and greater regulation. Whereas New Labour's third way accepted capitalism and globalization, they still perceived environmentalism as suggesting that capitalist enterprise is incompatible with sustainable development. Such perceptions were deeply unfair to the New environmental pragmatists who had embraced environmental modernization and worked with private corporations to promote environmental and social responsibility.

Jacobs suggests that environmental modernization offers New Labour a discourse that resonates with the public and policies to strengthen the government's delivery of ecological sustainability so that it matches progress

(continued)

(continued)

on economic and social sustainability. It allows the party to throw off green ideology and mainstream the environment within its centre-left programme in the same way that social democratic parties in Europe have done. Such modernization would capture the trends and mood of the times by fostering debate and action on the future of industrial society, quality of life, social inclusion and the nature and role of the state. It should:

- Go with the grain of globalization. Pick up and promote the trends towards higher environmental productivity and adapt the new knowledge-based economy towards environmental ends.
- Acknowledge individualisation and consumption but seek to encourage consumption towards environmentally benign forms. Collective consumption can contribute as much to the overall quality of life as individual.
- Give a central place to the perception of risk and scientific uncertainty and make risk-management a key policy field.
- Seek to counter the trend towards greater environmental inequality and exclusion.
- See the future as essentially optimistic with environmental problems as soluble (firmly a modernist project, accepting the central role of science and technology in tackling as well as contributing towards environmental problems). (Jacobs, 1999, p. 29).

A Better Quality of Life (Figure 13.4) provides a coherent account of new Labour's environmental commitments and allows an assessment of the extent to which it quietly embraced environmental modernization. Chapter 6 on a sustainable economy outlines the ways in which the government was encouraging greater resource efficiency by working with the market (targets, indicators, promotion of good practice, information for customers, funding of research and development) and only using regulation and taxation 'where appropriate'. It claims that market transformation will offer more people better goods and services at lower cost with lower environmental impact. With regard to the private car, this entails voluntary agreements with major manufactures on fuel efficiency and emissions; a sustainability strategy for motor manufacturers and traders; a cleaner-vehicle taskforce; an escalator on fuel charges (an annual 6 per cent increase in fuel duty); reform of company car taxation; and measures to cut journeys such as the Are You Doing Your Bit? campaign. The government's critics suggested that such policies were contradicted by others. Examples include the government's efforts to drive down car prices in the UK, its failure to support a 'car free' day throughout Europe; its reluctance to introduce congestion charges in cities; its continuing investment in road building; and the failure of ministers to be seen walking or cycling. These contradictions came to a head in the summer of 2000 as road hauliers, farmers and motorists protested over fuel price increases, largely caused by rises in the world price of oil. The protests seriously dented the Labour government's popularity and pointed to its continuing failure to embrace, justify and popularize policies of ecological modernization.

Environmental modernization is a necessary but not sufficient condition for realizing strong sustainability. Efficiency enthusiasts often confuse the macrolevel with the microlevel, inferring that good practice in one business can spread to the economy as a whole. But a lack of strong regulation, co-operation and planning, coupled with the imperatives of accumulation, prevents this. Under competitive conditions there are incentives to convert savings of money, resources or time into expanded output or use (e.g. mobile phones, paperless offices, more efficient heating systems). Demand may grow independently of higher unit efficiency and so offset resource and energy gains (e.g. cars, computers), while the dynamics of expansion at home and abroad continually increases ecological footprints (e.g. digital televisions left on standby, air-conditioning units replacing natural ventilation). Kovel (1999) reviews the arguments of green capitalists and suggests that the failure of environmental modernization will eventually force a profound accumulation crisis and forms of **ecofascism** as states become more authoritarian. He maintains that there can be no resolution of the ecological crisis so long as capital stands, that the programme of the modernizers cannot be realized through reform, and that only democratic socialism can realize sustainability.

The post-Rio agenda gives attention to consultation, participation and empowerment with decision-making at the most local level possible, and to partnerships and collaborations between all sectors with transparency, information, participation and access to justice (Figure 13.4). By fostering a more **radical and discursive democracy** it raises the possibilities of stronger forms of sustainability with greater attention to the conservation and development of ecological, human, and social and organizational capital. Environmental modernization may be linked to reflexive modernization to embrace green rationalism (Figure 1.8, p. 20), and the resulting reflection and action may lead to green socialist societies beyond industrialism and capitalism. These will redefine both the ends and means of life and use the opportunities provided by ecological limits and new technologies to liberate people from the treadmill of capitalist production and consumption (Pepper, 1993; 1996). Less work in the formal sector, a guaranteed income for all, more free time, reduced consumption, greater self-determination, real choice and coordinated socioeconomic planning across all levels of a cosmopolitan democracy to ensure equitable and sustainable distribution of global resources, are the keys to **post-industrial socialism** (Little, 1998). Based on a philosophy of self-limitation that recognizes ecological limits, it requires us to rethink modern notions of efficiency; reshape our needs, aspirations and narratives; and realize new kinds of wealth in selective slowness, more localized economies, and the reconstruction of public spheres (Sachs, 1999).

Gorz (1989; 1994), Bowring (1995), Goldblatt (1996) and Leipitz (1992) argue the case for such green socialism and the associated sustainability in the development mode is best seen in those experiments with community-based economies, politics and knowledge that are attempting to defend 'nature' as an environment that sustains the culture of everyday life (p. 5). **New economies of time and nature** and **new forms of welfare** lie at the heart of positive and sustainable forms of postmodernity that moderate modernity's obsession with growth, individualism, technocracy, consumerism and reductionism with renewed attention to sufficiency, collectivism, self-management, personal development and holism.

13.6 Local sustainability and the renewal of local democracy

Agenda 21 (Figure 13.4) singles out local government as having a special role in realizing sustainable development:

> Local authorities construct, operate and maintain economic, social and environmental infrastructure, oversee planning processes, establish local environmental policies and regulations, and assist in implementing national and subnational environmental policies. As the level of governance closest to the people, they play a vital role in educating, mobilizing and responding to the public to promote sustainable development (Chapter 28).

Two-thirds of the actions in *Agenda 21* require the active involvement of local authorities, and Chapter 28 calls on them to initiate Local Agenda 21 (LA21) processes or local partnerships for sustainable development. Around three-quarters of UK local authorities were involved in such processes by 1997 and the new Labour government expected all authorities to have them in place by 2000. Chapter 11 made reference to LA21 in the context of sustainable cities, developments in UK local governance, and the more diverse local environmental policy regimes introduced as part of new modes of local regulation.

At its best the LA21 approach to environmental planning (Figure 13.6) allows local communities radically to rethink how they live and make decisions. It involves local business, government and communities in establishing shared agendas, developing capacity for change, and learning through action to establish more sustainable localities and neighbourhoods. It has prompted many UK local authorities to widen and deepen their environmental agendas to embrace the economic, social and health dimensions of sustainability,

---

- Multisectoral engagement in the planning process, through local stakeholder groups, which serve as the coordination and policy bodies for preparing a long-term sustainable development action plan.
- Consultation with community groups, NGOs, business, churches, government agencies, professional groups and unions in order to create a shared vision and to identify proposals and priorities for action.
- Participatory assessment of local social, economic and environmental conditions and needs.
- Participatory target-setting through negotiations among key stakeholders in order to achieve the vision and goals set out in the action plan.
- Monitoring and reporting procedures, including local indicators, to track progress and allow participants to hold each other accountable to the action plan.

Figure 13.6 The distinguishing characteristics of the LA21 approach
Source: ICLEI, 1997.

Figure 13.7   Residents of Hitchin discuss sustainable futures, part of Hertfordshire's LA21 strategy

Source: Hertfordshire County Council.

and to experiment with new forms of policy-integration, public participation and community education (Buckingham-Hatfield and Percy, 1999; Selman, 1996; Young, 1997). LA21 requires and develops new (postmodern) approaches to environmental planning and public participation (Warburton, 1998) (Figure 13.7), and case studies of good practice (LGMB, 1997; 1998) suggest that it can foster radical democracy and the kind of praxis that allows stronger forms of sustainability to take root and grow. Geographers are among those who have studied these new approaches (e.g. Rydin and Pennington, 2000; Box 13.2).

In stressing local needs, socially useful production, popular planning, local economic strategies and environmental justice, some LA21 practice shows continuities with that of the new urban left of the 1980s (p. 203). The events surrounding the election of Ken Livingstone as mayor of London in 2000 suggest that this has continuing support but that its version of the '**third way**', as a form of democratic and libertarian socialism beyond social democracy and state collectivism, conflicts with that of a government that had diluted Labour's commitments to equality and democracy in favour of efficiency, pragmatism and partnership with the market (Cohen, 1999; Monbiot, 2000).

By valuing citizens, their knowledge, participation and action, LA21 has the potential to revive democracy and citizenship at a time when they are threatened by apathy and distrust. A study of public perceptions of sustainability in Lancashire revealed a pervasive sense of distrust and cynicism towards public institutions (MacNaghten *et al.*, 1995). National and local governments and politicians were seen to be unresponsive to people's needs, geared to short-term self-interest, beyond influence and unlikely to act on those social and environmental trends that provoked much anxiety and pessimism. While the majority had never heard of sustainable development, people identified with the concept once it was explained, particularly if it was related to the locality and such everyday concerns as jobs, health, crime and litter. Many people's continuing affectionate attachment to place, locality and community can be developed through LA21 to create the **human and social capital** that is at the core of citizenship and resists trends towards individualization and the breakdown of local networks, communities and identities. The accumulation of social and organizational capital rather than manufactured capital lies at the heart of strong sustainability and new roles for civil society and communitarianism (Etzioni, 2000; Tam, 1998) within local governance can work to advantage.

While there is much good LA21 practice, the process is often ill-defined and delivered by authorities that are underfunded, structurally constrained and further hindered by inadequate regional and national frameworks to coordinate their initiatives. There remain uncertainties over ends and means, and Marvin and Guy (1997) warn against a new localism that overlooks wider social trends and realities. LA21 processes and products often duplicate other local authority initiatives and/or fail to integrate with mainstream local governance. Hierarchy and departmentalism within councils delay a shift to more corporate and holistic approaches, and there is often a lack of political commitment from elected members and involvement by local business. LA21 working groups and round tables can remain marginal to the local political process and participation and consultation can remain substitutes for a lack of real action. The restructuring of local government (Chapter 11) has resulted in a real loss of functions, powers and resources, and unless national government is prepared to remove fiscal constraints, devolve more powers, and establish statutory mechanisms for delivering local policies, the LA21 process is likely to result in much disillusionment (Evans and Percy, 1999). Compulsory competitive tendering is one example of a restructuring initiative that acted against sustainability by precluding environmentally and socially responsible production and consumption, and it remains to be seen whether its replacement by best value will prove to be a more sustainable alternative.

13.7 The role of geography and geographers

LA21 and other progress towards sustainability suggests that individuals do play key roles. They act as catalysts for social change by living and advocating alternatives; as community champions by expressing the needs, visions and demands of their neighbours; and as networkers by developing links with others across space and between movements.

We hope that after reading this book you will be more prepared to adopt such roles and work for sustainability as a consumer, carer, worker and citizen.

As far as being a geographer is concerned we hope we have convinced you that you have a special responsibility and opportunity. Geographers continue to add to our understanding of the necessary transition to sustainability (Box 13.2) and by acting as **transformative intellectuals** can help people to see themselves and their relationship with the rest of the natural world in more critical and liberating ways. They can correct partial or false understanding, enable people to make good sense out of common sense, and help them glimpse the opportunities that ecological limits and sustainable development afford. You may not be able to do this directly as a teacher as we have sought to do in this text, but however you apply the kind of critical geography we have outlined, you should find many opportunities to contribute to a more sustainable world. The rewards are considerable and we hope you will respond to the challenge.

Box 13.2   Geographers researching and writing on sustainable development

Here are the abstracts from four articles written by geographers working in Britain in the 1990s. They provide pointers to geography's contribution to sustainable development and the kind of literature that allows readers of this text to develop their studies further. Books and articles by geographers from other parts of the world can be found in most university libraries.

Michael Redclift (1992), 'The Meaning of Sustainable Development'

The discussion of sustainable development has frequently proved confusing. Some writers are concerned with the sustainability of the natural resource base, others with present or future levels of production and consumption. Similarly, there are marked differences of opinion over the way in which sustainable development might be achieved. We need to examine the different dimensions of sustainability separately, to consider the kinds of international policies that would be required to achieve them, and the extent to which global solutions are either possible or available. This paper addresses these issues of theory and divergent intellectual tradition, and considers the problems and possibilities of reaching agreement between the countries of the North and those of the South.

Susan Owens (1994), 'Land, Limits and Sustainability: a conceptual framework and some dilemmas for the planning system' (see also Owens, 1997)

This paper explores the opportunities and contradictions in applying concepts of sustainable development to land-use policy. The conceptual framework is provided by 'stock maintenance' models of sustainability and a distinction is made between material, postmaterial and non-instrumental dimensions of sustainability which relate in complex ways to the use and development of land. Though

(continued)

(continued)

concepts of sustainability are gaining ground in planning and related disciplines, translating theory into policy remains problematic. Principles of sustainability challenge the presumption in favour of development and sit uneasily with the utilitarian notion of 'balance'. They require an alternative ethical basis and, especially in the postmaterial realm, are inherently bound up with value theory. These issues are illustrated by the problem of defining 'critical natural capital'. Political commitments to sustainability were made and to some extent encoded in planning policies, before the challenge to a demand-led economy was fully grasped. Far from effecting reconciliation, defining what is sustainable will expose conflict more starkly and at an earlier stage in the planning process. As environment-led plans and decisions are challenged by development interests, there will be opportunities to test these conclusions in specific empirical contexts.

Richard Munton (1997), 'Engaging Sustainable Development: some observations on progress in the UK'

Continuing criticism of the term 'sustainable development' deflects attention from the political project that has to underpin shifts in human behaviour. It is more important to redefine the roles and functions of public and private institutions in ways that legitimate sustainable actions in the belief that latent public support for environmental protection can be converted into a widening range of new but customary behaviours. A brief review of the UK government's response to the Rio declaration demonstrates faltering purpose, reinforcing and reflecting public distrust of government and those commercial interests that benefit from the status quo. Only very modest comfort can be taken from its limited initiatives which range from the greater dissemination of information to the setting of some environmental targets. The unwillingness of government to promote major fiscal or financial reforms, or significantly decentralize power and initiative, or to recognize the limitations of scientific knowledge when seeking the bases of 'rational' decision-making in the face of uncertainty, reveal the inadequacy of the steps so far taken.

Yvonne Rydin and Mark Pennington (2000), 'Public Participation and Local Environmental Planning: the collective action problem and the potential of social capital'

Expanding the opportunities for public participation in environmental planning is not always the best option. Starting from an institutional public choice analysis of public participation in terms of the collective action problem, this paper emphasizes the roots of participatory activities in the incentive structures facing potential participants. It then goes on to consider the strategies that may be adopted for encouraging greater public involvement and looks particularly at the social capital literature for suggestions of how institutional redesign may alter these incentive structures. The paper concludes by distinguishing three modes of environmental planning, in terms of the rationale of participation, the severity of the collective action problem and the associated participatory strategy that can be adopted.

13.8 Further reading

Read the report of the Commission on Global Governance (CGG, 1995). What progress has been made on implementing the Commission's proposals?

By reference to Connelly and Smith (1999, Chapter 8), Alder and Wilkinson (1999, pp. 86–93) and/or other sources, find out about European institutions and their role in promoting sustainable development.

Suggest further articles for Box 13.2 based on your reading of recent articles in the university library.

13.8.1 Websites

Earth Council for Sustainable Development: http://www.ecouncil.ac.cs/
Virtual Library of Sustainability: http://www.ulb.ac.be/ceese/sustvl.html
UNED–UK: http://www.oneworld.org/uned-uk/
International Institute for Sustainable Development: http://iisd1.iisd.ca/ic/
Forum for the Future: http://www.greenfutures.org.ok

Chapter 14

Envisioning an ecological socialist future

People have a need to experience themselves as part of an unfinished story that defines the past and present state of affairs and projects a future. Such a future should overcome the problems they currently face, should be one they can help to create, and should provide goals and aspirations to guide their everyday lives. **Utopias** fulfil this function, by questioning the existing social order and offering a dream that wants to be realized (Box 6.2, p. 107). They have inspired heroic efforts to create a better world, but postmodernization calls such grand narratives into question. Fewer people believe in social alternatives; that the world can be a better place; or that there is something preferable to the delights of consumer capitalism. They lack a totalizing perspective and consequently fail to understand themselves (Gare, 2000).

We hope that this book has encouraged you to think of yourself as an agent within a systemic and dialectical world in which there are new opportunities for humanity to co-evolve sustainably alongside the rest of nature. Advances in science and philosophy are undermining the mechanistic cosmology on which capitalism is based, while experiments with new kinds of technology, economy, work, community, welfare and governance, suggest ways in which we can constrain the market and democratize structures of power. We hope we have convinced you that geographers, who have long concerned themselves with the ecological and social relations shaping space and environments, now have new stories to tell. Geographical education can direct attention to sustainable futures with a new confidence, offering individuals and communities utopias as part of the process of social learning or praxis that lies at the heart of democratic change.

Different utopias are appropriate for different communities. We offer you one that we have used with students and community groups in Bedford as an example (Box 14.1). Illustrated by Brick's cartoons, (Clarke, 1990), it is designed to prompt reflection and action on alternatives beyond a reformed or managed capitalism as Gare argues:

> [it] is necessary to provide a compelling alternative to the consumerism of the affluent to which most people in the world now aspire, and justify this alternative. It is necessary to justify, affirm and celebrate the value of human creativity, sociality, sensitivity and cultural life, and beyond this, of all life, practically and theoretically in a way barely imaginable, at least under normal circumstances, within a capitalist regime (Gare, 2000, p. 29).

Box 14.1 Bedford 2045

It is a Wednesday in September 2045 and Jane Pearson wakes early. The sun is shining down on Bedford and the Pearson family's house in Queens Park. The solar collector on the roof has warmed the water for Jane's shower and by the time she has dressed and gone downstairs, husband Tom is giving Jake his breakfast. Before sitting down with them Jane takes some empty bottles and packaging out to the bins in the backyard. There are different bins for paper, glass, plastic, metals and organic waste, and their contents will be collected by the neighbourhood recycling centre that afternoon. Levels of household waste have fallen steeply since Bedford launched its local Agenda

21 in 1996. There are community composting schemes and organic allotments; most bulky packaging is returned to shops for recycling; and more goods are made in ways which allow them to be repaired and upgraded.

Jane, Jake and Tom tuck in to their breakfast of cereals and fresh fruit from the neighbourhood orchards. A lot of food is now grown around the town and Tom spends some of his time working at a local nursery where the glasshouses are heated with hot water from a small combined heat and power generating station which burns straw and willow. The rural landscape around Bedford has changed radically in the last 50 years. Farming is less intensive and more organic. There are smaller fields, a greater variety of crops and animals, and a great increase in semi-natural habitats and biodiversity. The Community Forest in the Marston Vale is now well established and there has also been extensive tree-planting in other areas and in the town itself. Most people do some work on the land and appreciate their contact with nature.

(continued)

(continued)

A 'greener' town means cleaner air and water, less noise and a kinder atmosphere in which to live.

Over breakfast Jane and Tom talk about their plans to add another room to their house before January when their second child will be born. Friends in the street will help them with some of the work once the prefabricated timber sections are delivered and they will engage a plumber and electrician through the town's local economic trading scheme (LETS), which now accounts for 30 per cent of local business turnover. They will need to get a low-interest loan from the Credit Union. Two years ago it helped Jane set up a small bicycle-repair workshop, and it will probably help again.

Tom and Jake leave home together as usual and Jake goes as far as the tram stop in the child's seat on the back of Tom's bike. Trams were introduced into Bedford in 1998 and now provide fast, clean, free public transport from west to east across the town, via the river embankment, and around the old inner ring road. Most people now live near enough to walk or cycle to work, but there are electric bus services and some light-rail links to surrounding towns which accommodate dual rail–road vehicles. Staggered working hours, more home working, and road-pricing for private cars, mean that traffic jams are a thing of the past. Bedford's high-speed rail links with the rest of the country have improved since the railways were privatized and people use the airship port at Cardington for occasional long-distance flights.

At the tram stop Tom meets his father Bill who is disabled and needs the tram to get him to the community centre where he helps look after young children like Jake. Tom and Bill exchange views about last night's concert at the Corn Exchange, then the tram arrives and Jake joins granddad for the short journey to the Westbourne Centre.

It takes Tom another five minutes to reach the engineering factory where he works for 20 hours each week. The regional government now guarantees all adults between eighteen and fifty-five this amount of work, and with a

(continued)

(continued)

national minimum wage it is generally sufficient to meet their needs. They can do additional paid work but few opt to do so. Most prefer to use non-work time for education, leisure and voluntary work, and this means that there is less stress and fewer health problems. Tom and his workmates make engines for electric buses and are proud of the contribution they have made to the recovery of Britain's vehicle industry. New forms of industrial relations and worker co-ownership mean that they have a real say in what is made and how it is produced, and they also take a share of the profits.

By ten o'clock Jane has arrived at work and is expecting a group from Biddenham Upper School. When they arrive she puts them to work repairing bicycles and takes the opportunity to tell them something about the way flywheels can store energy and how they can be incorporated into bicycle design. Cycle-repair, solar-collector installation and greenhouse construction, are all now part of the GCSE Practical Technology examination which the members of the group will be taking. They include not only teenagers but adults who join school classes as one way of meeting their educational and training needs.

Around three o'clock Jane finishes work and cycles to the community centre. Before collecting Jake she spends half an hour with an aerobics group and then joins Sue, John and Tasleem to work for an hour on the musical play they are writing which will be performed at the Riverside Arts Complex next year. It celebrates 50-years of Bedford's Local Agenda 21 and deals in a light-hearted way with changes in the local economy environment and ways of life. They work on a scene which deals with old and new ways of obtaining water and treating sewage. People now collect and use the water that falls on their roofs and driveways, and sewage is treated by a network of reed-beds beyond Priory Park. Water charges have fallen and the water companies now make much of their money by advising on conservation and pollution prevention.

(continued)

(continued)

On his way home Tom stops at the cobbler's shop to be measured for some shoes and at the neighbourhood repair centre to see if the camera is mended. The Pearsons are going to eat in the community café that evening so there is no need to shop for fresh food or to hurry home. Tom spends an hour at the cybercafé where he uses the Internet to join in a debate on LETS fraud and to send his views on the situation in Indonesia to his MEP and representative at the UN.

The community café, like the community laundry, is a way of sharing domestic work and saving energy. Some people work in them for wages which are set by the Neighbourhood Council, but most people work in them to obtain services at a cheaper rate and meet their neighbours. All the talk over dinner this evening is about the community meeting at eight o'clock.

The people of Queens Park have regular community meetings with members of their Neighbourhood Council. The council sends representatives to the Bedford Borough Council which in turn sends representatives to the Regional Assembly. All people over sixteen who live in Queens Park are eligible to sit on the Council and its membership is decided by a computer which selects 20 names at random every four years. This method of selecting councillors means that most people take an interest in local affairs and that Neighbourhood and Borough Council meetings are generally crowded and lively. This represents a great advance on the situation prevailing 50 years ago.

There are three main items on the agenda for this evening's meeting: to decide a new use for the old brewery site alongside the river; to fix the level of the neighbourhood consumption tax for the coming year; and to consider the borough's plans to withdraw its subsidy from the local centre for sporting excellence. A biotechnology park, student housing and a virtual-reality palace, have all been proposed for the brewery site; the consumption tax will probably have to rise as fewer people are buying new consumer goods; and there

(continued)

(continued)

is less interest in commercial spectator sport as people have found other things to do with their leisure time. Tom and Jane join in the discussion and Jane is elected to a group of citizens and councillors who will consider the brewery-site proposals in more detail and bring back ideas to the next meeting. Jane likes the idea of a virtual-reality palace. It could further reduce people's desire to travel while educating them about the wider world.

It is 10.30 by the time Tom and Jane have walked home. There's just time to use the home media centre to check the e-mail, vote in a referendum on the future of the national lottery, and catch up with the financial news on Channel 48. It seems that 3M has made major losses in China and that the future of its European plants may be at risk. A Brazilian company is looking for somewhere to manufacture cancer-curing drugs in Britain. Jane wonders whether or not they could be persuaded to come to Bedford.

14.1 Further reading

Read the case for a new metanarrative in part four of Jencks (1996).

Frankel (1986) provides a survey of post-industrial utopians including a chapter on getting from here to there.

Find out what Giddens (1994) means by utopian realism. Does this book foster such realism?

References

Agarwal, A. (1998) 'False Predictions', *Down to Earth*, 7 (1).

Agarwal, A. and Narain, S. (1991) *Global Warming in an Unequal World: a case of environmental colonialism*, New Delhi: Centre for Science and Environment.

Agarwal, A., Narain, S. and Sharma, A. (1999) *Green Politics, Global Environmental Negotiations 1*, New Delhi: Centre for Science and Environment.

Alder, J. and Wilkinson, D. (1999) *Environmental Law and Ethics*, Basingstoke: Macmillan.

Altvater, E. (1998) 'Global Order and Nature', in R. Keil, D. V. J. Bell, P. Penz and L. Fawcett (eds), *Political Ecology: Global and Local*, London: Routledge.

Anderson, P. (1996) *Passages from Antiquity to Feudalism*, London: Verso.

Arhem, K. (1996) 'The Cosmic Food Web: human–nature relatedness in Northwest Amazonia', in P. Descola and G. Palsson (eds), *Nature and Society: anthropological perspectives*, London: Routledge.

Atkins, P., Simmons, I. and Roberts, B. (1998) *People, Land and Time: an historical introduction to the relations between landscape, culture and environment*, London: Edward Arnold.

Atkinson, A. (1991) *Principles of Political Ecology*, London: Belhaven Press.

Attfield, R. (1991) *The Ethics of Environmental Concern*, Georgia: University of Georgia Press.

Baird, V. (1999) 'Green Cities', *New Internationalist*, 7–10 June.

Baldwin, R., Scott, C. and Hood, C. (eds) (1998) *A Reader on Regulation*, Oxford: Oxford University Press.

Barnes, T. and Gregory, D. (1997) *Reading Human Geography: the poetics and politics of inquiry*, London: Edward Arnold.

Barratt Brown, M. (1974) *The Economics of Imperialism*, Harmondsworth: Penguin.

Barry, J. (1999) *Environment and Social Theory*, London: Routledge.

Bate, J. (2000) *The Song of the Earth*, London: Picador.

Baudrillard, J. (1998) *The Consumer Society: myths and structures*, London: Sage.

Beck, U. (1992) *Risk Society*, London: Sage.

Beck, U. (1998) 'The Cosmopolitan Manifesto', *New Statesman*, 20 March.

Beheny, M. and Rookwood, R. (1993) 'Planning the Sustainable City Region' in A. Blowers (ed.), *Planning for a Sustainable Environment*, London: Earthscan.

Bendixson, T. and Platt, J. (1992) *Milton Keynes: image and reality*, Cambridge: Granta.

Blackburn, S. (1996) *The Oxford Dictionary of Philosophy*, Oxford: Oxford University Press.

Blowers, A. (1984) *Something in the Air: corporate power and the environment*, London: Harper & Row.

Blowers, A. (ed.) (1993) *Planning for a Sustainable Environment*, London: Earthscan.

Blunden, J. and Curry, N. (1988) *A Future for our Countryside*, Oxford: Blackwell.

Blunden, J. and Turner, G. (1985) *Critical Countryside*, London: BBC.

Bocock, R. (1993) *Consumption*, London: Routledge.

Bolton, G. (1981) *Spoils and Spoilers: Australians make their environment 1788–1980*, London: Allen & Unwin.

Bord, R., Fisher, A. and O'Connor, R. (1998) 'Public Perceptions of Global Warming: United States and international perspectives', *Climate Research*, 11: 75–84.

Borger, J. (1999) 'Giant Killers. How Europe's eco-warriors humbled the mighty Monsanto', *Guardian*, 22 November.

Boserup, E. (1965) *The Conditions of Agricultural Growth*, London: Allen & Unwin.

Bouchery, P. (1996) 'The Relationship between Society and Nature among the Hani People of China', in B. Formoso (ed.), *The Link with Nature and Divine Mediations in Asia*, Providence: Berghahn Books.

Bowler, I. (1991) 'The Agricultural Pattern', in R. Johnston and V. Gardner (eds), *The Changing Geography of the UK*, London: Routledge.

Bowring, F. (1995) 'Andre Gorz: ecology, system and lifeworld', *Capitalism Nature Socialism*, 6 (4): 65–84.

Brazier, C. (ed.) (1989) 'A History of the World', *New Internationalist*, 196, June.

Britnell, R. (1990) 'Feudal Reaction after the Black Death in the Palatinate of Durham', *Past and Present*, 128: 28–47.

Brown, P. (1996) *Global Warming: can civilization survive?* London: Blandford.

Brown, P. (2000a) 'Sellafield Crisis', *Guardian*, 28 March.

Brown, P. (2000b) 'Could This Be the End for Sellafield?', *Guardian*, 25 March.

Buckingham-Hatfield, S. and Percy, S. (eds) (1999) *Constructing Local Environmental Agendas: people, places + participation*, London: Routledge.

Budiansky, S. (1996) *Nature's Keepers: the new science of nature management*, London: Phoenix.

Byrne, D. (1998) *Complexity Theory and the Social Sciences: an introduction*, London: Routledge.

Capra, F. (1982) *The Turning Point*, New York: Simon & Schuster.

Capra, F. (1997) *The Web of Life: a new synthesis of mind and matter*, London: Flamingo.

Carley, M. and Spapens, P. (1998) *Sharing the World: sustainable living and global equity in the 21st century*, London: Earthscan.

Carson, R. (1962) *Silent Spring*, London: Hamish Hamilton.

CFS (1996) *The State of Canada's Forests 1995–1996: sustaining forests at home and abroad*, Ottawa: Canadian Forest Service.

CFS (1999) *The State of Canada's Forests 1998–1999: innovation*, Ottawa: Canadian Forest Service.

CGG (Commission on Global Governance) (1995) *Our Global Neighbourhood*, Oxford: Oxford University Press.

Charlesworth, J. and Cochrane, A. (1994) 'Tragedy, Farce and the Postmodern City: the case of Milton Keynes', Occasional Paper Series No. 9, Faculty of Social Sciences, Open University, Milton Keynes.

Cho, G. (1995) *Trade, Aid and Global Interdependence*, London: Routledge.

Chomsky, N. (1996) *Powers and Prospects: reflections on human nature and the social order*, London: Pluto Press.

Clapp, B. (1994) *An Environmental History of Britain*, Harlow: Longman.

Clark, N., Brown, R., Parker, E., Robins, T., Remick, D., Philbert, M., Keeler, G. and Israel, B. (1999) 'Childhood Asthma', *Environmental Health Perspectives*, 107 (3): 421–9.

Clarke, J. (1990) The cartoons used to illustrate Box 14.1 were originally drawn by John Clarke (Brick) for a classroom activity in *Environment and Democracy (What We Consume, Unit 10)*, J. Huckle, Richmond, WWF/Richmond Publishing, 1990.

Clastres, P. (1998) *Chronicle of the Guayaki Indians*, transl. by Pool Awter, London: Faber & Faber.

Cloke, P., Philo, C. and Sadler, D. (1991) *Approaching Human Geography*, London: Paul Chopman.

Clutton-Brock, J. (1987) *Domesticated Animals from Early Times*, Cambridge: Cambridge University Press.

Cohan, S. and Shires, L. (1988) *Telling Stories: a theoretical analysis of narrative fiction*, London: Routledge.

Cohen, N. (1999) *Cruel Britannia*, London: Verso.

Collier, A. (1994) *Critical Realism: an introduction to Roy Bhaskar's philosophy*, London: Verso.

Connelly, J. and Smith, G. (1999) *Politics and the Environment: from theory to practice*, London: Routledge.

Cook, I. and Pepper, D. (eds) (1990) 'Anarchism and Geography', *Contemporary Issues in Geography and Education*, 3 (2).

Coote, S. (1995) *William Morris: his life and work*, Oxford: Past Times.

Corbridge, S. and Agnew, J. (1995) *Mastering Space: hegemony, territory and international political economy*, London: Routledge.

Cornforth, M. (1987) *Materialism and the Dialectical Method*, London: Lawrence & Wishart.

Cosgrove, D. (1990) 'Environmental Thought and Action: premodern and postmodern', *Transactions of the Institute of British Geographers*, NS, 15: 344–58.

Counsell, D. (1999a) 'Making Sustainable Development Operational', 'Environmental Protection Approaches Within Structure Plans' and 'Accommodating Social Needs' (a series of three articles), *Town and Country Planning*, 68 (4–6): 131–3, 159–61, 202–4.

Counsell, D. (1999b) 'Attitudes to Sustainable Development in Planning: policy integration, participation and Local Agenda 21, a case-study of the Hertfordshire Structure Plan', *Local Environment*, 4 (1): 21–32.

Coveney, R. and Highfield, R. (1995) *Frontiers in Complexity: the search for order in a chaotic world*, London: Faber & Faber.

Cowie, J. (1998) *Climate and Human Change: disaster or opportunity*, Carnforth: Parthenon.

Craig, A. (2000) 'The Green Man', *New Statesman*, 10 July: 56–7.

Crang, P. (1999) 'Local–Global', in P. Cloke, P. Crang and M. Goodwin (eds), *Introducing Human Geographies*, London: Edward Arnold.

Croall, S. and Rankin, W. (2000) *Introducing Environmental Politics*, London: Icon Books.

Crockatt, R. (1995) *The Fifty Years War: the United States and the Soviet Union in world politics, 1941–1991*, London: Routledge.

Crosby, A. (1986) *Ecological Imperialism: the biological expansion of Europe, 900–1900*, Cambridge: Cambridge University Press.

Crow, B. and Thomas, A. (1994) *Third World Atlas*, Milton Keynes: Open University Press.

Cullingworth, B. (1996) 'A Vision Lost', *Town and Country Planning*, 65 (6): 172–4.

Daly, H. and Cobb, J. (1990) *For the Common Good: redirecting the economy towards community, the environment and a sustainable future*, London: Green Print.

David, D. (1997) 'Globalization: some key questions', *Courier*, 164: 50–54.

Deb, D. (1993) 'Evolution of Joint Forest Management in South-West Bengal: a subaltern detail', *Social Action*, 43 (3): 368–74.

Devall, B. and Sessions, G. (1999) 'Deep Ecology', in M. Smith (ed.), *Thinking Through the Environment*, London: Routledge.

Dickens, P. (1992) *Society and Nature: towards a green social theory*, London: Harvester Wheatsheaf.

Dickens, P. (1996) *Reconstructing Nature: alienation, emancipation and the division of labour*, London: Routledge.

Dobson, A. (1995) *Green Political Thought*, London: Routledge.

Dodds, F. (ed.) (1997) *The Way Forward, Beyond Agenda 21*, London: Earthscan.

Dodds, K. (2000) *Geopolitics in a Changing World*, Harlow: Prentice Hall.

Dodgshon, R. (1990) 'The Changing Evaluation of Space 1500–1914, in R. Dodgshon and R. Butlin (eds), *An Historical Geography of England and Wales*, 2nd edn, London: Academic Press.

Dodgshon, R. and Butlin, R. (eds) (1990) *An Historical Geography of England and Wales*, 2nd edn, London: Academic Press.

DoE (Department of the Environment) (1996) *Sustainable Settlements and Shelter: the United Kingdom national report habitat II*, London: HMSO.

Doyle, T. and McEachern, D. (1998) *Environment and Politics*, London: Routledge.

Drabble, M. (1984) *A Writer's Britain*, London: Thames & Hudson.

Dryzek, J. (1996) 'Political and Ecological Communication', in F. Matthews (ed.), *Ecology and Democracy*, London: Frank Cass.

Dryzek, J. (1997) *The Politics of the Earth*, Oxford: Oxford University Press.

DTER (Department of Transport, Environment and the Regions) (1999) *A Better Quality of Life*, London: HMSO.

Dyck, R. (1998) 'Integrating Planning and Sustainability Theory for Local Benefit', *Local Environment*, 3 (1): 27–42.

Eden, S. (1998) 'Environmental Knowledge, Uncertainty and the Environment', *Progress in Human Geography*, 22 (3): 425–32.

Ehrlich, P. (1972) *The Population Bomb*, London: Ballantine.

Ehrlich, P. and Ehrlich, A. (1998) *Betrayal of Science: how anti-environment rhetoric threatens our future*, Washington, DC: Island Press.

Ekins, P., Hillman, M. and Hutchison, R. (1992) *Wealth Beyond Measure: an atlas of the new economics*, London: Gaia Books.

Elliott, L. (1998) *The Global Politics of the Environment*, Basingstoke: Macmillan.

Ellwood, W. (ed.) (2000) 'Redesigning the Global Economy', *New Internationalist*, 320, January/February.

Elwin, V. (1939) 'Civilising the Savage', in R. Guha (ed.) (1994) *Social Ecology*, Delhi: Oxford University Press.

Elwin, V. (1958) *Leaves from the Jungle: life in a Gond village*, 2nd edn, Delhi: Oxford University Press.

Engels, F. (1950) 'Introduction to Dialectics of Nature' and 'The Part Played by Nature in the Transition from Ape to Man', in K. Marx and F. Engels, *Selected Works, Vol. 2*, London: Lawrence & Wishart.

Environmental News Service (ENS) (1999) 'Carbon Dioxide Emissions Drop, Despite Delay in Kyoto Protocol', http://www.ens.lycos.com

Etzioni, A. (2000) *The Third Way to the Good Society*, London: Demos.

Evans, B. and Percy, S. (1999) 'The Opportunities and Challenges for Local Environmental Policy and Action in the UK', in S. Buckingham-Hatfield and S. Percy (eds), *Constructing Local Environmental Agendas: people, places and participation*, London: Routledge.

Evernden, N. (1992) *The Social Creation of Nature*, Baltimore, MD: Johns Hopkins University Press.

Fernandes, W. (1987) 'Afforestation Programmes, Voluntary Action and Community Organisation', *Social Action*, 37: 275–95.

Feuer, L. (ed.) (1978) *Marx and Engels, Basic Writings on Politics and Philosophy*, London: Fontana/Collins.

Food and Drink Federation (1999) *Food for our Future*, London.

Forest Alliance (2000) 'Interfor takes FSC Challenge', *Sustainability Update: a quarterly publication of the Forest Alliance*, spring 2000.

Formoso, B. (1996) 'Tai Cosmology and the influence of Buddhism', in B. Formoso (ed.), *The Link with Nature and Divine Mediations in Asia*, Providence: Berghahn Books.

Frankel, B. (1986) *The Post-Industrial Utopians*, Cambridge: Polity Press.

Friends of Clayoquot Sound (2000a) 'Is there Hope in the Biosphere Reserve?' news release, 2 February.

Friends of Clayoquot Sound (2000b) 'Interfor Logging Park Border Sparks Action by Clayoquot Biosphere Protectors', news release, 29 May.

FSC (2000) 'Principles and Criteria', Document 1.2 (amended February 2000), Forest Stewardship Council, Oaxaca, www.fscoax.org

Fuller, P. (1985) *Images of God: The consolation of lost illusions*, London: Tigerstripe.

Gaarder, J. (1995) *Sophie's World*, London: Phoenix.

Gadgil, M. and Guha, R. (1992) *This Fissured Land: an ecological history of India*, Delhi: Oxford University Press.

Gadotti, M. (1996) *Pedagogy of Praxis: a dialectical philosophy of education*, New York: SUNY.

Gamble, A. (1981) *An Introduction to Modern Social and Political Thought*, Basingstoke: Macmillan.

Gandy, M. (1996) 'Crumbling Land: the postmodernity debate and the analysis of environmental problems', *Progress in Human Geography*, 20 (1): 23–40.

Gare, A. (2000) 'Creating an Ecological Socialist Future', *Capitalism Nature Socialism*, 11 (2): 23–40.

Georgescu-Roegen, N. (1971) *The Entropy Law and the Economic Process*, Cambridge, MA: Harvard University Press.

Ghatak, S. (1995) *Introduction to Development Economics*, London: Routledge.

Gibbs, D. (1996) 'Integrating Sustainable Development and Economic Restructuring: a role for regulation theory', *Geoforum*, 27 (1): 1–10.

Giddens, A. (1984) *The Constitution of Society*, Cambridge: Polity Press.

Giddens, A. (1989) *Sociology*, Cambridge: Polity Press.

Giddens, A. (1994) *Beyond Left and Right: the future of radical politics*, Cambridge: Polity Press.

Giddens, A. (1998) 'After the Left's Paralysis', *New Statesman*, 1 May: 18–21.

Giddens, A. (1999) 'Globalisation', Reith Lecture No. 1, London: BBC.

Giradet, H. (1996) *Gaia Atlas to Cities*, London: Gaia.

Goldblatt, D. (1996) *Social Theory and the Environment*, Cambridge: Polity Press.

Goldman, M. (1992) 'Environmentalism and Nationalism', in J. Massey Stewart (ed.), *The Soviet Environment: problems, policies and politics*, Cambridge: Cambridge University Press.

Goldsmith, E., Allen, R., Allaby, M., Davoll, J. and Lawrence, S. (1972) *Blueprint for Survival*, Harmondsworth: Penguin.

Goodwin, M. and Painter, J. (1996) 'Local Governance, the Crises of Fordism and the Changing Geographies of Regulation', *Transactions of the Institute of British Geographers*, 21: 635–48.

Gorz, A. (1989) *Critique of Economic Reason*, London: Verso.

Gorz, A. (1993) 'Political Ecology: expertocracy versus self-limitation', New Left Review, 202: 55–67.

Gorz, A. (1994) *Capitalism, Socialism, Ecology*, London: Verso.

Gottlieb, R. (1995) 'Spiritual Deep Ecology and the Left: an effort at reconciliation', *Capitalism Nature Socialism*, 6 (3): 1–21.

Goudie, A. (1990) *The Human Impact on the Natural Environment*, Oxford: Basil Blackwell.

Goudie, A. (1993) *The Human Impact: man's role in environmental change*, Oxford: Basil Blackwell.

Goudie, A. and Viles, H. (1997) *The Earth Transformed: an introduction to human impacts on the environment*, Oxford: Blackwell.

Government of British Columbia (2000) 'British Columbia Community Celebrates Designation of Clayoquot Sound as an International Biosphere Reserve', News Release, 5 May.

Government of India (1952) *National Forest Policy*, New Delhi.

Government of India (1988) *National Forestry Policy*, New Delhi.

Government of India (1990) *Involvement of Village Communities and Voluntary Agencies in the Regeneration of Degraded Forests*, Circular No. 6.21/89 – FP, 1 June, New Delhi.

Graham, K. (2000) *The Planetary Interest*, London: UCL Press.

Gray, J. (1998) *False Dawn: the delusions of global capitalism*, London: Granta.

Greenpeace (1998) *British Columbia Communities at the Crossroads: towards ecological and economic sustainability*, Vancouver: Greenpeace.

Grigg, D. (1995) *An Introduction to Agricultural Geography*, 2nd edn, London: Routledge.

Grimes, D. (1995) *Like Dew Before the Sun: life and language in Northamptonshire*, Wellingborough: Wharton.

Grove-White, R., Macnaughten, P., Mayer, S. and Wynne, B. (1997) *Uncertain World, Genetically Modified Organisms, Food and Public Attitudes in Britain*, Lancaster: Centre for Study of Environmental Change.

Guha, R. (1983) 'Forestry in British and Post-British India: an historical analysis', *Economic & Political Weekly*, 18 (44): 1882–1896.

Guha, R. (1995) 'Mahatma Gandhi and the Environmental Movement in India', *Capitalism Nature Socialism*, 6 (3): 47–61.

Gyford, J. (1985) *The Politics of Local Socialism*, London: Allen & Unwin.

Hadjor, K. (1993) *Dictionary of Third World Terms*, Harmondsworth: Penguin.

Hall, P. (1996) '1946–1996: from New Town to Sustainable Social City', *Town and Country Planning*, 65 (11): 295–7.

Hall, P. (1998) 'The Three Magnets Re-interpreted', *Town and Country Planning*, 67 (10): 6–8.

Hallo, W. and Simpson, W. (1971) *The Ancient Near East: a history*, New York: Harcourt Brace Jovanovich.

Halpin, M. (1981) *Totem Poles: an illustrated guide*, Vancouver: University of British Columbia Press.

Hardin, G. (1968) 'The Tragedy of the Commons', *Science*, 162: 1243–48.

Hardy, Y. (1997) 'Sustainable Forest Management: the mark of a society', Ottawa: Canadian Forest Service.

Harrison, C. and Burgess, J. (1994) 'Social Constructions of Nature: a case-study of conflicts over the development of Rainham Marshes', *Transactions of the Institute of British Geographers*, 19: 291–310.

Hartmann, F. (1998) 'Towards a Social Ecological Politics of Sustainability', in R. Keil, D. Bell, P. Penz and L. Fawcett (eds), *Political Ecology: global and local*, London: Routledge.

Harvey, D. (1974) 'Population, Resources, and the Ideology of Science', *Economic Geography*, 50: 256–77.

Harvey, D. (1989a) *The Condition of Postmodernity: an enquiry into the origins of cultural change*, Oxford: Blackwell.

Harvey, D. (1989b) *The Urban Experience*, Oxford: Blackwell.

Harvey, D. (1996) *Justice, Nature and the Geography of Difference*, Oxford: Blackwell.

Haughton, G. and Hunter, C. (1996) *Sustainable Cities*, London: Jessica Kingsley/ Regional Studies Association.

Heater, D. (1974) *Contemporary Political Ideas*, Harlow: Longman.

Heathcote, R. (1975) *Australia*, Harlow: Longman.

Hebbert, M. (1998) 'Tomorrow Never Came – or Did it?', *Town and Country Planning*, 67 (10): 17–19.

Hecht, S. and Cockburn, A. (1989) *The Fate of the Forest: developers, destroyers and defenders of the Amazon*, London: Verso.

Heise, J. (1996) *Akkadian Language*, www.sron.ruu.nl/~jheise/akkadian/

Held, D. (1995) *Democracy and the Global Order: from the modern state to cosmopolitan governance*, Cambridge: Polity Press.

Hetherington, P. (2000a) 'Town Grouse and Country Grouse', *Guardian*, 16 February.

Hetherington, P. (2000b) 'Building Blocks', *Guardian*, 22 May.

Hinchcliffe, S. (1999) 'Cities and Nature: intimate strangers', in J. Allen, D. Massey and M. Pryke (eds), *Unsettling Cities*, London: Routledge.

Hindess, B. and Hirst, P. (1975) *Pre-capitalist Modes of Production*, London: Routledge & Kegan Paul.

Hines, C. (2000) *Localization, a Global Manifesto*, London: Earthscan.

HMSO (1975) *Social Trends*, London: HMSO.

HMSO (1980) *Social Trends*, London: HMSO.

HMSO (1987) *Social Trends*, London: HMSO.

HMSO (1990) *Social Trends*, London: HMSO.

HMSO (1995) *Social Trends*, London: HMSO.

Hobsbawm, E. (1997) *The Age of Revolution, The Age of Capital, The Age of Empire* (3 vols), London: Abacus.

Holt-Jensen, A. (1988) *Geography, History and Concepts*, London: Paul Chapman.

Hopkins, T. and Wallerstein, I. (eds) (1996) *The Age of Transition: trajectory of the world-system 1945–2025*, London: Zed Books.

Hoskins, W. (1955) *The Making of the English Landscape*, London: Book Club Associates.

Hough, M. (1995) *Cities and Natural Process*, London: Routledge.

Houghton, J. (1994) *Global Warming: the complete briefing*, Oxford: Lion Publishing.

Howard, E. (1889) *Tomorrow: a peaceful path to real reform*, revi. edn (1965), entitled *Garden Cities of Tomorrow*, London: Faber & Faber.

Howe, L. M. (2000) 'Hamoukar, Syria – a city older than 6,000 years?', Earthfiles, www.earthfiles.com

Hughes, J. (1994) 'Ecology in Ancient Civilisations', in D. Ward (ed.), *Green History*, London: Routledge.

Huntley, B. (1990) 'Lessons from Climates in the Past', in J. Leggett, *Global Warming: the Greenpeace Report*, Oxford: Oxford University Press.

IBRAD (1992) *Problems and Prospects of Participatory Community Development*, Working Paper No. 15, Calcutta: Indian Institute of Biosocial Research and Development.

ICLEI (International Council for Local Environmental Initiatives) (1997) *Local Agenda 21 Survey: a study of responses by local authorities and their national and international associations to Agenda 21*, Toronto: ICLEI.

Ikeda, S. (1996) 'World Production', in T. Hopkins and I. Wallerstein (eds), *The Age of Transition: trajectory of the world-system 1945–2025*, London: Zed Books.

IPCC (1996a) *The Economic and Social Dimensions of Climate Change: summary for policy-makers*, Geneva: IPCC Working Group III.

IPCC (1996b) *The Science of Climate Change: summary for policy-makers*, Geneva: IPCC Working Group I.

IPCC (1996c) *Second Assessment Synthesis of Scientific–Technical Information Relevant to Interpreting Article 2 of the UN Framework Convention on Climate Change*, Geneva: IPCC.

IPCC (1997a) *Stabilization of Atmospheric Greenhouse Gases: physical, biological and socioeconomic implication*, February 1997 Technical Paper, Geneva: IPCC Working Group I.

IPCC (1997b) *The Regional Impacts of Climate Change: an assessment of vulnerability: summary for policymakers*, November 1997 Special Report, Geneva: IPCC Working Group II.

Irwin, A. (1995) *Citizen Science*, London: Routledge.

IUCN/UNEP/WWF (1991) *Caring for the Earth: A Strategy for Sustainable Living*, Gland: International Union for the Conservation of Nature.

Jacobs, J. (1961) *The Death and Life of Great American Cities*, Harmondsworth: Penguin.

Jacobs, M. (1996) *The Politics of the Real World*, London: Earthscan.

Jacobs, M. (1997) 'Introduction: The New Politics of the Environment' in M. Jacobs (ed.), *Greening the Millennium? The New Politics of the Environment, Political Quarterly*, Oxford: Blackwell.

Jacobs, M. (1999) *Environmental Modernization: the New Labour agenda*, London: Fabian Society.

Jacobs, M. (ed.) (1997) *Greening the Millennium? The new politics of the environment, Political Quarterly*, Oxford: Blackwell.

Jäger, J. and O'Riordan, T. (eds) (1996) *Politics of Climate Change: a European perspective*, London: Routledge.

Jeffery, R. and Sundar, N. (1999) *A New Moral Economy for India's Forests? Discourses of community and participation*, New Delhi: Sage.

Jencks, C. (1996) *What is Post-Modernism?*, London: Academy Editions.

Jodha, N. (1986) 'Common Property Resources and Rural Poor in Dry Regions of India', *Economic & Political Weekly*, 21 (27): 1169–1181.

Johnston, R. (1989a) *Philosophy and Human Geography: an introduction to contemporary approaches*, London: Edward Arnold.

Johnston, R. (1989b) *Environmental Problems: nature, economy and state*, London: Belhaven.

Jones, B., Gray, A., Kavanagh, D., Moran, M., Norton, P. and Seldon, A. (1991) *Politics UK*, London: Philip Allen.

Keen, M. (1991) 'Robin Hood: a peasant hero', *History Today*, October: 20–24.

Keil, R. and Graham, J. (1998) 'Reasserting Nature: constructing urban environments after Fordism', in B. Braun and N. Castree (eds), *Remaking Reality: nature at the millenium*, London: Routledge.

Kellner, D. (1989) *Critical Theory, Marxism and Modernity*, Baltimore, MD: Johns Hopkins University Press.

Kelly, P. (1999) 'The Geographies and Politics of Globalization', *Progress in Human Geography*, 23 (3): 379–400.

Kipfer, S., Hartmann, F. and Marino, S. (1996) 'Cities, Nature and Socialism: towards an urban agenda for action and research', *Capitalism Nature Socialism*, 7 (2): 5–20.

Klein, N. (2000) *No Logo*, London: Flamingo.

Kovel, J. (1999) 'The Justifiers: a critique of Julian Simon, Stephan Schemidheiny, and Paul Hawken on capitalism and nature, *Capitalism Nature Socialism*, 10 (3): 3–37.

Krausskopf, G. (1996) 'Nepalese Chiefs and Gods', in B. Formoso (ed.), *The Link with Nature and Divine Mediations in Asia*, Providence: Berghahn Books.

Kumar, D. (1992) 'Eucalypts in Industrial and Social Plantations in Karnataka', in I. Calder, R. Hall and P. Adlard (eds), *Growth and Water Use of Forest Plantations*, Chichester: John Wiley.

Kurlansky, M. (1999) *Cod*, London: Vintage.

Lane, D. (1985) *Soviet Economy and Society*, Oxford: Basil Blackwell.

Lash, S., Szerszynski, B. and Wynne, B. (eds) (1996) *Risk, Environment and Modernity: towards a new ecology*, London: Sage.

Lash, S. and Urry, J. (1994) *Economies of Signs and Space*, London: Sage.

Lecompte-Tilouine, M. (1996) 'The Cult of the Earth Goddess among the Magar of Nepal', in B. Formoso (ed.), *The Link with Nature and Divine Mediations in Asia*, Providence: Berghahn Books.

Lee, R. and Daly, R. (1999) *The Cambridge Encyclopædia of Hunters and Gatherers*, Cambridge, Cambridge: University Press.

Lee, R. and de Vore, I. (eds) (1968) *Man the Hunter*, Chicago: Aldine.

Leipitz, A. (1992) *Towards a New Economic Order: postfordism, ecology and democracy*, Cambridge: Polity Press.

Lekachman, R. and van Loon, B. (1981) *Capitalism for Beginners*, London: Writers and Readers.

Lewin, R. (1997) *Complexity: life on the edge of chaos*, London: Phoenix.

LGMB (Local Government Management Board) (1995) *Indicators for Local Agenda 21 – a summary*, Luton: LGMB.

LGMB (Local Government Management Board) (1997, 1998) *Local Agenda 21 Case Studies* (two folders), London: LGMB.

Libert, B. (1995) *The Environmental Heritage of Soviet Agriculture*, Wallingford: CAB International.

Litfin, K. (1994) *Ozone Discourses: science and politics in global environmental cooperation*, New York: Columbia University Press.

Little, A. (1998) *Post-industrial Socialism: towards a new politics of welfare*, London: Routledge.

Livingstone, D. (1992) *The Geographical Tradition*, Oxford: Blackwell.

Loh, J., Randers, J., MacGillivray, A., Kapos, V., Jenkins, M., Groombridge, B. and Cox, N. (1998) *Living Planet Report*, Gland: Worldwide Fund for Nature.

Lovins, A., Lovins, H. and von Wiezacker, E. (1997), *Factor Four: doubling wealth, halving resource use*, London: Earthscan.

Lyon, D. (1999) *Postmodernity*, 2nd edn, Milton Keynes: Open University Press.

Maclure, M. (1995) 'Postmodernism: a postscript', *Educational Action Research*, 3 (1): 105–16.

MacNaghten, P., Grove-White, R., Jacobs, M. and Wynne, B. (1995) *Public Perceptions and Sustainability in Lancashire: indicators, institutions, and participation*, Preston: Lancashire County Council.

MacNaghten, P. and Urry, J. (1998) *Contested Natures*, London: Sage.

Marsden, T., Murdoch, J., Lowe, P., Munton, R. and Flynn, A. (1993) *Constructing the Countryside*, London: UCL Press.

Martell, L. (1994) *Ecology and Society: an introduction*, Cambridge: Polity Press.

Marvin, S. and Guy, S. (1997) 'Creating Myths rather than Sustainability: the transition fallacies of the new localism', *Local Environment*, 2 (3): 311–18.

Marx, K. (1975) 'Estranged Labour', in *Early Writings*, Harmondsworth: Penguin Books.

Mason, M. (1999) *Environmental Democracy*, London: Earthscan.

Matthews, F. (1991) *The Ecological Self*, London: Routledge.

McKay, G. (1996) *Senseless Acts of Beauty: cultures of resistance since the sixties*, London: Verso.

McKibben, W. (1990) *The Death of Nature*, Harmondsworth: Penguin.

Mellaart, J. (1975) *The Neolithic of the Near East*, London: Thames & Hudson.

Merchant, C. (1992) *Radical Ecology: the search for a liveable world*, London: Routledge.

Micklin, P. (1992) 'Water Management in Soviet Central Asia', in J. Massey Stewart (ed.), *The Soviet Environment: problems, policies and politics*, Cambridge: Cambridge University Press.

Miller, G. (1995) *Environmental Science: working with the earth*, Belmont: Wadsworth.

Monbiot, G. (2000) *Captive State*, London: Pan Macmillan.

Moorehead, A. (1987) *The Fatal Impact: the invasion of the South Pacific 1767–1840*, Sydney: Mead & Beckett.

Morris, A. (1994) *History of Urban Form: before the industrial revolution*, Harlow: Longman.

Morton, A. (ed.) (1979) *Political Writings of William Morris*, London: Lawrence & Wishart.

Mumford, L. (1961) *The City in History*, Hardmondsworth: Penguin.

Munton, R. (1997) 'Engaging Sustainable Development: some observations on progress in the UK', *Progress in Human Geography*, 21 (2): 147–63.

NASA (1999) Global Temperature Trends, NASA Goddard Institute for Space Studies, http://www.giss.nasa.gov/research/observe/surftemp/

Naughton, J. (1999) *A Brief History of the Future: the origins of the internet*, London: Weidenfeld & Nicolson.

O'Connor, J. (1988) 'Capitalism, Nature, Socialism: a theoretical introduction', *Capitalism Nature Socialism*, 1 (1): 1–19.

O'Donnell, R., Mitchell, P., Priest, N., Strange, L., Fox, A., Henshaw, D. and Long, S. (1997) 'Variations in the Concentration of Plutonium, Strontium-90 and Total Alpha-emitters in Human Teeth Collected in the British Isles', *Science of the Total Environment*, 201: 235–43.

O'Leary, B. (1989) *The Asian Mode of Production*, Oxford: Basil Blackwell.

O'Riordan, T. (1981) *Environmentalism*, London: Pion.

O'Riordan, T. (1996) *EcoTaxation*, London: Earthscan.

O'Riordan, T. and Jordan, A. (1997) 'Kyoto in Perspective', *ECOS*, 18 (3/4): 38–42.

O'Riordan, T. and Jordan, A. (1999) 'Institutions, Climate Change and Cultural Theory: towards a common analytical framework', *Global Environmental Change*, 9 (2): 81–93.

O'Tuathail, G., Dalby, S. and Routledge, P. (1998) *The Geopolitics Reader*, London: Routledge.

Odum, E. (1997) *Ecology, a Bridge between Science and Society*, Sunderland: Sinauer Associates.

Onimode, B. (1985) *An Introduction to Marxist Political Economy*, London: Zed Books.

Osbourne, R. and Edney, (1992) *Philosophy for Beginners*, New York: Writers and Readers.

Ostrom, E. (1990) *Governing the Commons: the evolution of institutions for collective action*, Cambridge: Cambridge University Press.

Owen, L. and Unwin, T. (eds) (1997) *Environmental Management: readings and case studies*, Oxford: Blackwell.

Owens, S. (1994) 'Land, Limits and Sustainability: a conceptual framework and some dilemmas for the planning system', *Transactions of the Institute of British Geographers*, 19: 439–56.

Owens, S. (1997) 'Interpreting Sustainable Development: the case of land-use planning', in M. Jacobs (ed.), Greening the Millennium? The new politics of the environment, *Political Quarterly*, Oxford: Blackwell.

Paterson, M. (1996) *Global Warming and Global Politics*, London: Routledge.

Patterson, A. and Theobald, K. (1999) 'Emerging Contradictions; sustainable development and the new local governance', in S. Buckingham-Hatfield and S. Percy (eds), *Constructing Local Environmental Agendas: people, places and participation*, London: Routledge.

Peet, R. (1989) 'World Capitalism and the Destruction of Regional Cultures', in R. Johnston and P. Taylor (eds), *The World in Crisis*, 2nd edn, Oxford: Blackwell.

Peet, R. (1991) *Global Capitalism: theories of societal development*, London: Routledge.

Peet, R. (1998) *Modern Geographical Thought*, Oxford: Blackwell.

Pennington, W. (1969) *The History of British Vegetation*, London: English Universities Press.

Pepper, D. (1984) *The Roots of Modern Environmentalism*, London: Routledge.

Pepper, D. (1993) *Eco-socialism, from Deep Ecology to Social Justice*, London: Routledge.

Pepper, D. (1995) 'Misrepresenting Deep Ecology and the Left', *Capitalism Nature Socialism*, 6 (3): 39–41.

Pepper, D. (1996) *Modern Environmentalism: an introduction*, London: Routledge.

Percy, M. (1996) 'The best-laid plans', *Town and Country Planning*, 65 (3): 75–82.

Pereira, W. and Seabrook, J. (1990) *Asking the Earth: farms, forestry and survival in India*, London: Earthscan.

Pilkington, A. (1997) 'Is Sociology Dead? Sociology, the Enlightenment and the challenge of the postmodernists', *Sociology Review*, 6 (8): 22–5.

Pinchot, G. (1999) 'Conservation and Human Welfare', in Smith, M. (ed.), *Thinking through the Environment*, London: Routledge/Open University.

Plumwood, V. (1993) *Feminism and the Mastery of Nature*, London: Routledge.

Ponting, C. (1991) *A Green History of the World*, London: Sinclair-Stevenson.

Porritt, J. (2000) *Playing Safe: science and the environment*, London: Thames & Hudson.

Porter, P. and Sheppard, E. (1998) *A World of Difference: society, nature, development*, New York: Guilford Press.

Postan, M. (1972) *Agrarian Life in the Middle Ages*, Harmondsworth: Penguin.

Powell, D., Leiss, W., Griffiths, A. and Barrett, K. (1997) 'Gene Escape, or the Pall of Silence over Plant Biotechnology Risk', in D. Powell and W. Leiss (eds), *Mad Cows and Mother's Milk: the perils of poor risk communication*, London: McGill/Queen's University Press.

Powell, J. (1998) *Postmodernism for Beginners*, London: Writers and Readers.

Powell, J. M. (1991) *An Historical Geography of Modern Australia: the restive fringe*, Cambridge: Cambridge University Press.

Pratt, V. with Howarth, J. and Brady, E. (2000) *Environment and Philosophy*, London: Routledge.

Pretty, J. (1990) 'Sustainable Agriculture in the Middle Ages: the English manor', *Agricultural History Review*, 38 (1): 1–19.

Pretty, J. (1998) 'Social Capital and Sustainable Livelihoods', paper presented at 15th International Symposium of the Association for Farming Systems Research-Extension, Pretoria.

Proctor, J. (1998) 'The Social Construction of Nature: relativist accusations, pragmatist and realist responses', *Annals of the Association of American Geographers*, 88 (3): 352–76.

Pryce, R. (1977) 'Approaches to the Study of Man and Environment', *Man and Environment, Section 1, Unit 2, D204 Fundamentals of Human Geography*, Milton Keynes: Open University Press.

Quaini, M. (1982) *Geography and Marxism*, Oxford: Basil Blackwell.

Rackham, O. (1994) *The Illustrated History of the British Countryside*, London: Weidenfeld & Nicolson.

Rahman, A. (1993) *People's Self-Development*, London: Zed Books.

Rahnema, M. and Bawtree, V. (1997) *The Post-Development Reader*, London: Zed Books.

Redclift, M. (1984) *Development and the Environmental Crisis: red or green alternatives*, London: Routledge.

Redclift, M. (1987) *Sustainable Development: exploring the contradictions*, London: Routledge.

Redclift, M. (1992) 'The Meaning of Sustainable Development', *Geoforum*, 23 (3): 395–403.

Reed, M. (1990) *The Landscape of Britain: from the beginning to 1914*, London: Routledge.

Rees, J. (1985) *Natural Resources, Allocation, Economics and Policy*, London: Routledge.

Rees, W. (1997) 'Is "Sustainable City" an Oxymoron?', *Local Environment*, 2 (3): 303–10.

Reid, D. (1995) *Sustainable Development: an introductory guide*, London: Earthscan.

Reifer, T. and Sidler, J. (1996) 'The Interstate System', in T. Hopkins and I. Wallerstein (eds), *The Age of Transition: trajectory of the world-system 1945–2025*, London: Zed Books.

Reynolds, S. (1994) *Fief and Vassals*, Oxford: Oxford University Press.

Rhodes, T. (1999) 'Bitter Harvest: the real story of Monsanto and GM food', *Sunday Times*, 22 August.

Richards, P. (1985) *Indigenous Agricultural Revolution: ecology and food production in West Africa*, London: Unwin Hyman.

Rifkin, J. (1999a) 'The Perils of the Biotech Century', *New Statesman*, 6 September: 12–13.

Rifkin, J. (1999b) *The Biotech Century: how genetic commerce will change the world*, Smyrna, GA: Phoenix Press.

Rifkin, J. and Howard, T. (1985) *Entropy: a new world-view*, London: Paladin.

Robins, N. (1999) 'Taming World Trade', *Resurgence*, 197: 32–4.

Robinson, G. (1990) *Conflict and Change in the Countryside*, London: Belhaven Press.

Roche, M. (1992) *Rethinking Citizenship: welfare, ideology and change in modern society*, Cambridge: Polity Press.

Rogers, R. and Gumuchdjian, P. (1997) *Cities for a Small Planet*, London: Faber & Faber.

Rowell, A. (1996) *Green Backlash: global subversion of the environment movement*, London: Routledge.

Roy, A. (1999) *The Cost of Living*, London: Flamingo.

Rydin, Y. (1999) 'Environmental Governance for Sustainable Urban Development: a European model?', *Local Environment*, 4 (1): 61–5.

Rydin, Y. and Pennington, M. (2000) 'Public Participation and Local Environmental Planning: the collective action problem and the potential of social capital', *Local Environment*, 5 (2): 153–69.

Sachs, W. (1999) *Planet Dialectics: explorations in environment and development*, London: Zed Books.

Sahlins, M. (1972) *Stone Age Economics*, London: Tavistock.

Saldanha, I. (1996) 'Colonialism and Professionalism: a German forester in India', *Economic and Political Weekly*, 31 (21): 1265–73.

Sardar, Z. (2000) 'A Science for Us All, Not Just for Business', *New Statesman*, 7 February: 30–31.

Sarin, M. (1995) 'Joint Forest Management in India: achievements and unaddressed challenges', *Unasylva*, 46: 30–36.

Satterthwaite, D. (1999) *Sustainable Cities: an Earthscan reader*, London: Earthscan.

Saunders, J. (1999) 'Seizing the Reins', *New Internationalist*, June: 28–9.

Saxena, N. C. (1993) 'Sustainable Development of Forest Lands and Joint Forest Management in India', *Indian Journal of Public Administration*, 39 (3): 465–72.

Sayer, A. (1985) 'Realism in Geography' in R. J. Johnston (ed.), *The Future of Geography*, London: Methuen.

Schama, S. (1995) *Landscape and Memory*, London: Fontana Press.

Schumacher, E. F. (1973) *Small is Beautiful: economics as if people mattered*, London: Abacus.

Scott, A., Berkhout, F. and Scoones, I. (eds) (1999) *The Politics of GM food: risk, science and public trust*, Lancaster: Centre for the Study of Environmental Change/ESRC.

Selman, P. (1996) *Local Sustainability: Managing and Planning Ecologically Sound Places*, London: Paul Chapman.

Shaw, A. (1960) *The Story of Australia*, London: Faber & Faber.

Shiva, V. (1993) 'The Greening of Global Reach', in W. Sachs (ed.), *Global Ecology: a new arena of political conflict*, London: Zed Books.

Shiva, V. (1998) *Biopiracy: the plunder of nature and knowledge*, London: Green Books.

Shoard, M. (1987) *This Land is Our Land: the struggle for Britain's countryside*, London: Paladin.

Short, J. R. (1992) *Imagined Country: Society, Culture and Environment*, London: Routledge.

Shostak, M. (1981) *Nisa: the life and words of a !Kung woman*, New York: Random House.

Simmons, C. and Chambers, N. (1998) 'Footprinting UK Households: how big is your ecological garden?', *Local Environment*, 3 (3): 355–62.

Simmons, I. (1989) *Changing the Face of the Earth: culture, environment, history*, Oxford: Blackwell.

Simmons, I. (1993) *Interpreting Nature: cultural constructions of the environment*, Oxford: Blackwell.

Simmons, I. (1997) *Humanity and Environment: a cultural ecology*, Harlow: Longman.

Simon, J. (1998) *The Ultimate Resource 2*, Princeton, NJ: Princeton University Press.

Slater, D. (1997) *Consumer Culture and Modernity*, Cambridge: Polity Press.

Slaughter, C. (1985) *Marx and Marxism: an introduction*, Harlow: Longman.

Smith, M. (1998) *Ecologism: towards ecological citizenship*, Million Keynes: Open University Press.

Smith, M. (ed.) (1999) *Thinking through the Environment: a reader*, London: Routledge/Open University Press.

Smith, N. (1984) *Uneven Development: nature, capital and the production of space*, Oxford: Basil Blackwell.

Smith, N. (1996) 'The Production of Nature', in G. Robertson, M. Mash, L. Tickner, J. Bird, B. Curtis and T. Putnam (eds), *Future Natural: nature, science, culture*, London: Routledge.

Smith, N. (1998) 'Nature at the Millenium: production and re-enchantment', in B. Braun and N. Castree (eds), *Remaking Reality: nature at the millenium*, London: Routledge.

Snaiberg, A. (1980) *The Environment: from surplus to scarcity*, Oxford: Oxford University Press.

Soper, K. (1995) *What is Nature? Culture, politics and the non-human*, Oxford: Blackwell.

Soper, K. (1999) 'The Politics of Nature: reflections on hedonism, progress and ecology', *Capitalism Nature Socialism*, 10 (2): 47–70.

SPWD (1992) *Joint Forest Management: concept and opportunity, proceedings of the national workshop at Surajkund*, August, New Delhi: Society for Promotion of Wasteland Development.

SPWD (1993) *Joint Forest Management Update*, New Delhi: Society for Promotion of Wasteland Development.

Sundar, N. (2000) 'Unpacking the "Joint" in Joint Forest Management', *Development and Change*, 31: 255–79.

Swift, R. (2000) 'Democracy: is that all there is?', *New Internationalist*, 324: 9–12.

Swyngedouw, E. (1996) 'The City as a Hybrid: on nature, society and cyborg urbanization', *Capitalism Nature Socialism*, 7 (2): 65–80.

Szerszynski, B., Lash, S. and Wynne, B. (1996) 'Ecology, Realism and the Social Sciences', in S. Lash, B. Szerszynski and B. Wynne (eds), *Risk, Environment and Modernity: towards a new ecology*, London: Sage.

Tam, H. (1998) *Communitarianism: a new agenda for politics and citizenship*, London: Macmillan.

Taylor, J. (1997) 'Hot Air in Kyoto', Cato Institute, http://www.cato.org/dailys/12-08-97.html

Taylor, P. (1989) *Political Geography: world-economy, nation-state and locality*, Harlow: Longman.

Tendler, J. (1997) *Good Government in the Tropics*, Baltimore, MD: Johns Hopkins University Press.

Thomas, K. (1984) *Man and the Natural World: changing attitudes in England 1500–1800*, Harmondsworth: Penguin.

Thoreau, H. D. (2000) *Wild Fruits*, New York: Norton.

Todaro, M. (1996) *Economic Development*, Harlow: Longman.

Toub, L. (1997) 'Monsanto's empire', *New Internationalist*, 293: 16.

Trainer, T. (1998) 'Towards a Checklist for Ecovillage Development', *Local Environment*, 3 (1): 79–84.

Tuan, Y.-F. (1974) *Topophilia: a study of environmental perception, attitudes and values*, London: Prentice Hall.

UNCED (1992a) Report of the United Nations Conference on Environment and Development: Annex II, *Agenda 21*, A/CONF.151/26 (vols 1–3), 12 August.

UNCED (1992b) Report of the United Nations Conference on Environment and Development: Annex I, *Rio Declaration on Environment and Development*, A/CONF. 151/26 (vol. I), 12 August.

UNDP (United Nations Development Programme) (1998) *Human Development Report*, New York: Oxford University Press.

UNEP (United Nations Environment Programme) (1999) *Global Environmental Outlook 2*, Nairobi: UNEP.

van der Post, L. (1958) *The Lost World of the Kalahari*, London: Hogarth Press.

Vidal, J. (1999) 'The Seeds of Wrath', *Guardian Weekend*, 19 June.

Visvanathan, S. (1991) 'Mrs Brundtland's Disenchanted Cosmos', *Alternatives*, 16 (2).

von Molthe, K. and Rahman, A. (1996) 'External Perspectives on Climate Change: a view from the United States and the Third World', in J. Jäger and T. O'Riordan (eds), *Politics of Climate Change: a European perspective*, London: Routledge.

Wackernagel, M. and Rees, W. (1996) *Our Ecological Footprint: reducing human impact on the earth*, Gabriola Island: New Society Publishers.

Wagner, R., Flynn, J., Gregory, R., Mertz, C. and Slovic, P. (1998) 'Acceptable Practices in Ontario's Forests: differences between the public and forestry professionals', *New Forests*, 16 (2): 139–54.

Walker, D. (1982) *The Architecture and Planning of Milton Keynes*, London: The Architectural Press.

Walker, P. (1998) 'Politics of Nature: an overview of political ecology', *Capitalism Nature Socialism*, 9 (1): 131–44.

Wallace, I. (1990) *The Global Economic System*, London: Unwin Hyman.

Wallerstein, I. (1984) *The Politics of the World Economy*, Cambridge: Cambridge University Press.

Warburton, D. (ed.) (1998) *Community and Sustainable Development: participation in the future*, London: Earthscan.

Ward, D. (1994) *Green History: a reader in environmental literature, philosophy and politics*, London: Routledge.

Watson, R. (1999) 'Report to the 10th Session of the Subsidiary Body for Scientific and Technical Advice on the status of the IPCC', Chairman's Report, 31 May 99, IPCC, Geneva.

WCED (World Commission on Environment and Development) (1987) *Our Common Future*, Oxford: Oxford University Press.

Westoby, J. (1989) *Introduction to World Forestry*, Oxford: Basil Blackwell.

White, R. (1994) *Urban Environmental Management*, Chichester: John Wiley.

Wigley, T., Richels, R. and Edmonds, J. (1996) 'Economic and Environmental Choices in the Stabilisation of Atmospheric CO_2 Concentrations', *Nature*, 379: 240–43.

Williams, H. (1988) *Whale Nation*, London: Cape.

Wilson, C. (1987) *Australia, the Creation of a Nation, 1788–1988*, London: Weidenfeld & Nicolson.

WIMSA (1998) 'Report on Activities, April 1997 to March 1998', Windhoek, Namibia: Working Group of Indigenous Minorities in Southern Africa.

Wintour, P. and Wainwright, M. (2000) 'Fresh Blow to Nuclear Plant', *The Guardian*, 27 March.

Wittfogel, K. (1957) *Oriental Despotism*, New Haven, CT: Yale University Press.

WMO (1999) 'World Meteorological Organization Annual Statement on the Global Climate', http://www.wmo.ch

Wolfson, Z. (1992) 'The Massive Degradation of Ecosystems in the USSR', in J. Massey Stewart (ed.), *The Soviet Environment: problems, policies and politics*, Cambridge: Cambridge University Press.

Wood, J. (1978) *Sun, Moon and Standing Stones*, Oxford: Oxford University Press.

World Resources Institute (1990) *World Resources 1990–1991*, Washington: WRI.

World Resources Institute (1998) *World Resources 1998–1999*, Washington: WRI.

Worster, D. (1992) *Rivers of Empire: water, aridity and the growth of the American West*, Oxford: Oxford University Press.

Wu, J. (1996) *Systems Dialectics*, Beijing: Foreign Language Press.

WWF (1999) 'America's Global Warming Solutions', www.worldwildlife.org/climate

Wynne, B. (1994) 'Scientific Knowledge and the Global Environment', in M. Redclift and T. Benton (eds), *Social Theory and the Global Environment*, London: Routledge.

Wynne, B. (1996) 'May the Sheep Safely Graze? A reflexive view of the expert–lay knowledge divide', in Lash *et al.* (eds), *Risk, Environment and Modernity: towards a new ecology*, London: Sage.

Wynne, B. (1999) 'Bitter Fruits: the issue of GM crops is too important to leave to science alone', *Guardian Science*, 16 September: 2–3.

Yearly, S. (1994) 'Social Movements and Environmental Change', in M. Redclift and T. Benton (eds), *Social Theory and the Global Environment*, London: Routledge.

Young, S. (1997) 'Local Agenda 21: the renewal of local democracy?', in M. Jacobs (ed.), *Greening the Millennium? The new politics of the environment, Political Quarterly*, Oxford: Blackwell.

Zepezaner, M. (1994) *The CIA's Greatest Hits*, New York: Adonian.

Ziegler, C. (1987) *Environmental Policy in the USSR*, London: Francis Pinter.

Zimmerer, K. (1994) 'Human Geography and the "New Ecology": the prospect and promise of integration', *Annals of the Association of American Geographers*, 84 (1): 108–25.

Index

Note: Pages numbers in *italics* refer to
Figures and Tables.

!Kung people 48–50

aboriginal people 98, 101
accommodatory solutions 136
accumulated capital 65
adaptation 59–60
administrative rationalism 21
aerosol particles 142
aesthetic appreciation of nature 163–4
agency and resistance 157–8
Agenda 21 205, 236, *237*, 242
agrarian capitalism 84–7
agricultural mode of production 50–3
   in British Isles 52–3
agriculture
   in feudal England 69–71
   postwar production, UK 119–21, *121*
   sedentary 43
AIDS 159
air pollution 152
alienation 33
alienation from nature 124
American Revolution 86
anarchism 106
Anglo–Iranian Oil Company 116
animism 26, 56–8
annual allowable cut (AAC) 181
anti-foundationalism 35, 39
anti-totalization 35–6, 39
anti-utopianism 36, 39
Aquinas, St Thomas 76
Aral Sea 128–30
Arbenz, Jacobo 116
Aristotle 26, 76
Armas, General Castillo 117

Association of Small Island States (AOSIS)
   149
Australia
   pastoralism in 98–101
   as semi-peripheral space 104
Australian Agricultural Company 99
autonomy, principle of 233–4

Bacon, Francis 27, 28
Baiga tribe of central India 58
Balfour Beatty 156
Ball, John 75
Baudrillard, Jean 36
Beck, Ulrich 159, 161
bewar 58
Bible 76, 77
biotechnology 154, 216–22
Black Act (1723) 89
Black Death 75
Blair, Tony 224
Blake, William 29
Bookchin, Murray 179
bourgeois revolution 79–81
bourgeoisie 80–1
Bretton Woods Agreement 115
brick-making 9–10
British Columbia, forestry in 176–83
British Energy 161
British Nuclear Fuels Limited (BNFL)
   160–1
Broecker salt conveyer *143*
bronze tools 62
Brown, Capability *30*
Brundtland, Gro Harlem 235
Brundtland Report: *Our Common Future*
   135, 235–6
Buddhism 77
Butterfly, Julia (Julia Hill) 175